HOLT®

ChemFile® B

Microscale Experiments

HOLT, RINEHART AND WINSTON

A Harcourt Education Company

Orlando • **Austin** • New York • San Diego • Toronto • London

Contents

Introduction to the Lab Program

Structure of the Experiments

INTRODUCTION

The opening paragraphs set the theme for the experiment and summarize its major concepts.

OBJECTIVES

Objectives highlight the key concepts to be learned in the experiment and emphasize the science process skills and techniques of scientific inquiry.

MATERIALS

These lists enable you to organize all apparatus and materials needed to perform the experiment. Knowing the concentrations of solutions is vital. You often need this information to perform calculations and to answer the questions at the end of the experiment.

SAFETY

Safety cautions are placed at the beginning of the experiment to alert you to procedures that may require special care. Before you begin, you should review the safety issues that apply to the experiment.

PROCEDURE

By following the procedures of an experiment, you perform concrete laboratory operations that duplicate the fact-gathering techniques used by professional chemists. You learn skills in the laboratory. The procedures tell you how and where to record observations and data.

DATA AND CALCULATIONS TABLES

The data that you collect during each experiment should be recorded in the labeled Data Tables provided. The entries you make in a Calculations Table emphasize the mathematical, physical, and chemical relationships that exist among the accumulated data. Both types of tables should help you to think logically and to formulate your conclusions about what occurs during the experiment.

CALCULATIONS

Space is provided for all computations based on the data you gather.

QUESTIONS

Based on the data and calculations, you should be able to develop plausible explanations for the phenomena you observe during the experiment. Specific questions require you to draw on the concepts you learn.

GENERAL CONCLUSIONS

This section asks broader questions that bring together the results and conclusions of the experiment and relate them to other situations.

Safety in the Chemistry Laboratory

CHEMICALS ARE NOT TOYS.

Any chemical can be dangerous if it is misused. Always follow the instructions for the experiment. Pay close attention to the safety notes. Do not do anything differently unless told to do so by your teacher.

Chemicals, even water, can cause harm. The trick is to know how to use chemicals correctly so that they will not cause harm. You can do this by following the rules on these pages, paying attention to your teacher's directions, and following the cautions on chemical labels and experiments.

These safety rules always apply in the lab.

1. Always wear a lab apron and safety goggles.

Even if you aren't working on an experiment, laboratories contain chemicals that can damage your clothing, so wear your apron and keep the strings of the apron tied. Because chemicals can cause eye damage, even blindness, you must wear safety goggles. If your safety goggles are uncomfortable or get clouded up, ask your teacher for help. Try lengthening the strap a bit, washing the goggles with soap and warm water, or using an antifog spray.

2. Generally, no contact lenses are allowed in the lab.

Even while wearing safety goggles, you can get chemicals between contact lenses and your eyes, and they can cause irreparable eye damage. If your doctor requires that you wear contact lenses instead of glasses, then you may need to wear special safety goggles in the lab. Ask your doctor or your teacher about them.

3. Never work alone in the laboratory.

You should always do lab work under the supervision of your teacher.

4. Wear the right clothing for lab work.

Necklaces, neckties, dangling jewelry, long hair, and loose clothing can cause you to knock things over or catch items on fire. Tuck in a necktie or take it off. Do not wear a necklace or other dangling jewelry, including hanging earrings. It isn't necessary, but it might be a good idea to remove your wristwatch so that it is not damaged by a chemical splash.

Pull back long hair, and tie it in place. Nylon and polyester fabrics burn and melt more readily than cotton, so wear cotton clothing if you can. It's best to wear fitted garments, but if your clothing is loose or baggy, tuck it in or tie it back so that it does not get in the way or catch on fire.

Wear shoes that will protect your feet from chemical spills—no open-toed shoes, sandals, or shoes made of woven leather straps. Shoes made of solid leather or a polymer are much better than shoes made of cloth. Also, wear pants, not shorts or skirts.

5. Only books and notebooks needed for the experiment should be in the lab.

Do not bring other textbooks, purses, bookbags, backpacks, or other items into the lab; keep these things in your desk or locker.

6. **Read the entire experiment before entering the lab.**
Memorize the safety precautions. Be familiar with the instructions for the experiment. Only materials and equipment authorized by your teacher should be used. When you do the lab work, follow the instructions and the safety precautions described in the directions for the experiment.

7. **Read chemical labels.**
Follow the instructions and safety precautions stated on the labels. Know the location of Material Safety Data Sheets for chemicals.

8. **Walk carefully in the lab.**
Sometimes you will carry chemicals from the supply station to your lab station. Avoid bumping other students and spilling the chemicals. Stay at your lab station at other times.

9. **Food, beverages, chewing gum, cosmetics, and tobacco are *never* allowed in the lab.**
You already know this.

10. **Never taste chemicals or touch them with your bare hands.**
Also, keep your hands away from your face and mouth while working, even if you are wearing gloves.

11. **Use a sparker to light a Bunsen burner.**
Do not use matches. Be sure that all gas valves are turned off and that all hot plates are turned off and unplugged before you leave the lab.

12. **Be careful with hot plates, Bunsen burners, and other heat sources.**
Keep your body and clothing away from flames. Do not touch a hot plate just after it has been turned off. It is probably hotter than you think. Use tongs to heat glassware, crucibles, and other things and to remove them from a hot plate, a drying oven, or the flame of a Bunsen burner.

13. **Do not use electrical equipment with frayed or twisted cords or wires.**

14. **Be sure your hands are dry before you use electrical equipment.**
Before plugging an electrical cord into a socket, be sure the electrical equipment is turned off. When you are finished with it, turn it off. Before you leave the lab, unplug it, but be sure to turn it off first.

15. **Do not let electrical cords dangle from work stations; dangling cords can cause tripping or electric shocks.**
The area under and around electrical equipment should be dry; cords should not lie in puddles of spilled liquid.

16. **Know fire drill procedures and the locations of exits.**

17. **Know the locations and operation of safety showers and eyewash stations.**

18. **If your clothes catch on fire, *walk* to the safety shower, stand under it, and turn it on.**

19. **If you get a chemical in your eyes, walk immediately to the eyewash station, turn it on, and lower your head so that your eyes are in the running water.**
Hold your eyelids open with your thumbs and fingers, and roll your eyeballs around. You have to flush your eyes continuously for at least 15 min. Call your teacher while you are doing this.

20. **If you have a spill on the floor or lab bench, don't try to clean it up by yourself.**
First, ask your teacher if it is OK for you to do the cleanup; if it is not, your teacher will know how the spill should be cleaned up safely.

21. **If you spill a chemical on your skin, wash it off under the sink faucet, and call your teacher.**
If you spill a solid chemical on your clothing, brush it off carefully so that you do not scatter it, and call your teacher. If you get a liquid chemical on your clothing, wash it off right away if you can get it under the sink faucet, and call your teacher. If the spill is on clothing that will not fit under the sink faucet, use the safety shower. Remove the affected clothing while under the shower, and call your teacher. (It may be temporarily embarrassing to remove your clothing in front of your class, but failing to flush that chemical off your skin could cause permanent damage.)

22. **The best way to prevent an accident is to stop it before it happens.**
If you have a close call, tell your teacher so that you and your teacher can find a way to prevent it from happening again. Otherwise, the next time, it could be a harmful accident instead of just a close call.

23. **All accidents should be reported to your teacher, no matter how minor.**
Also, if you get a headache, feel sick to your stomach, or feel dizzy, tell your teacher immediately.

24. **For all chemicals, take only what you need.**
On the other hand, if you do happen to take too much and have some left over, **do not** put it back in the bottle. If somebody accidentally puts a chemical into the wrong bottle, the next person to use it will have a contaminated sample. Ask your teacher what to do with any leftover chemicals.

25. ***Never* take any chemicals out of the lab.**
You should already know this rule.

26. **Horseplay and fooling around in the lab are very dangerous.**
Never be a clown in the laboratory.

27. **Keep your work area clean and tidy.**
After your work is done, clean your work area and all equipment.

28. **Always wash your hands with soap and water before you leave the lab.**

29. **Whether or not the lab instructions remind you, *all* of these rules *apply all of the time*.**

QUIZ

Determine which safety rules apply to the following.

- Tie back long hair, and confine loose clothing. (Rule ? applies.)
- Never reach across an open flame. (Rule ? applies.)
- Use proper procedures when lighting Bunsen burners. Turn off hot plates and Bunsen burners that are not in use. (Rule ? applies.)
- Be familiar with the procedures and know the safety precautions before you begin. (Rule ? applies.)
- Use tongs when heating containers. Never hold or touch containers with your hands while heating them. Always allow heated materials to cool before handling them. (Rule ? applies.)
- Turn off gas valves that are not in use. (Rule ? applies.)

SAFETY SYMBOLS

To highlight specific types of precautions, the following symbols are used in the experiments. Remember that no matter what safety symbols and instructions appear in each experiment, all of the 29 safety rules described previously should be followed at all times.

EYE AND CLOTHING PROTECTION

- Wear safety goggles in the laboratory at all times. Know how to use the eyewash station.

- Wear laboratory aprons in the laboratory. Keep the apron strings tied so that they do not dangle.

CHEMICAL SAFETY

- Never taste, eat, or swallow any chemicals in the laboratory. Do not eat or drink any food from laboratory containers. Beakers are not cups, and evaporating dishes are not bowls.
- Never return unused chemicals to their original containers.
- Some chemicals are harmful to the environment. You can help protect the environment by following the instructions for proper disposal.
- It helps to label the beakers and test tubes containing chemicals.
- Never transfer substances by sucking on a pipet or straw; use a suction bulb.
- Never place glassware, containers of chemicals, or anything else near the edges of a lab bench or table.

CAUSTIC SUBSTANCES

- If a chemical gets on your skin or clothing or in your eyes, rinse it immediately, and alert your teacher.
- If a chemical is spilled on the floor or lab bench, tell your teacher, but do not clean it up yourself unless your teacher says it is OK to do so.

HEATING SAFETY

- When heating a chemical in a test tube, always point the open end of the test tube away from yourself and other people.

EXPLOSION PRECAUTION

- Use flammable liquids in small amounts only.
- When working with flammable liquids, be sure that no one else in the lab is using a lit Bunsen burner or plans to use one. Make sure there are no other heat sources present.

HAND SAFETY

- Always wear gloves or use cloths to protect your hands when cutting, fire polishing, or bending hot glass tubing. Keep cloths clear of any flames.
- Never force glass tubing into rubber tubing, rubber stoppers, or corks. To protect your hands, wear heavy leather gloves or wrap toweling around the glass and the tubing, stopper, or cork, and gently push the glass tubing into the rubber or cork.
- Use tongs when heating test tubes. Never hold a test tube in your hand to heat it.
- Always allow hot glassware to cool before you handle it.

GLASSWARE SAFETY

- Check the condition of glassware before and after using it. Inform your teacher of any broken, chipped, or cracked glassware because it should not be used.
- Do not pick up broken glass with your bare hands. Place broken glass in a specially designated disposal container.

GAS PRECAUTION

- Do not inhale fumes directly. When instructed to smell a substance, waft it toward you. That is, use your hand to wave the fumes toward your nose. Inhale gently.

RADIATION PRECAUTION

- Always wear gloves when handling a radioactive source.
- Always wear safety goggles when performing experiments with radioactive materials.
- Always wash your hands and arms thoroughly after working with radioactive materials.

HYGIENIC CARE

- Keep your hands away from your face and mouth.
- Always wash your hands before leaving the laboratory.

Any time you see any of the safety symbols, you should remember that all 29 of the numbered laboratory rules always apply.

Labeling of Chemicals

In any science laboratory the labeling of chemical containers, reagent bottles, and equipment is essential for safe operations. Proper labeling can lower the potential for accidents that occur as a result of misuse. Read labels and equipment instructions several times before you use chemicals or equipment. Be sure that you are using the correct items, that you know how to use them, and that you are aware of any hazards or precautions associated with their use.

All chemical containers and reagent bottles should be labeled prominently and accurately with labeling materials that are not affected by chemicals.

Chemical labels should contain the following information:

1. **Name of the chemical and its chemical formula**

2. **Statement of possible hazards** This is indicated by the use of an appropriate signal word, such as *DANGER, WARNING,* or *CAUTION*. This signal word usually is accompanied by a word that indicates the type of hazard present, such as *POISON, CAUSES BURNS, EXPLOSIVE,* or *FLAMMABLE*. Note that this labeling should not take the place of reading the appropriate Material Safety Data Sheet for a chemical.

3. **Precautionary measures** Precautionary measures describe how users can avoid injury from the hazards listed on the label. Examples include: "use only with adequate ventilation" and "do not get in eyes or on skin or clothing."

4. **Instructions in case of contact or exposure** If accidental contact or exposure does occur, immediate first-aid measures can minimize injury. For example, the label on a bottle of acid should include this instruction: "In case of contact, flush with large amounts of water; for eyes, rinse freely with water for 15 min and get medical attention immediately."

5. **The date of preparation and the name of the person who prepared the chemical** This information is important for maintaining a safe chemical inventory.

Suggested Labeling Scheme	
Name of contents	hydrochloric acid
Chemical formula and concentration or physical state	6 M HCl
Statements of possible hazards and precautionary and measures	WARNING! CAUSTIC and CORROSIVE—CAUSES BURNS Avoid contact with skin and eyes Avoid breathing vapors.
Hazard Instructions for contact or overexposure	IN CASE OF CONTACT: Immediately flush skin or eyes with large amounts of water for at least 15 min; for eyes, get medical attention immediately!
Date prepared or obtained	May 8, 2005
Manufacturer (commercially obtained) or preparer (locally made)	Prepared by Betsy Byron, Faribault High School, Faribault, Minnesota

Laboratory Techniques

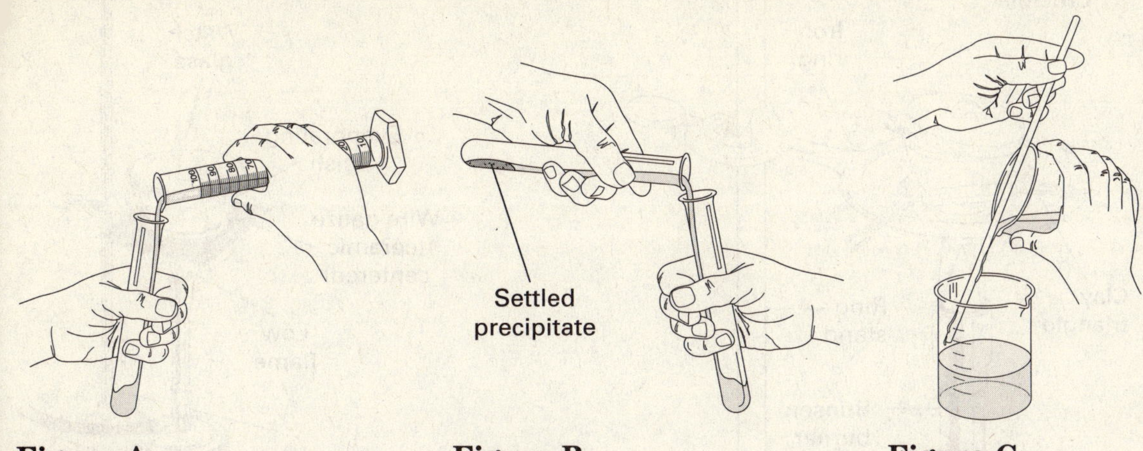

DECANTING AND TRANSFERRING LIQUIDS

1. The safest way to transfer a liquid from a graduated cylinder to a test tube is shown in **Figure A.** Transfer the liquid at arm's length with your elbows slightly bent. This position enables you to see what you are doing and still maintain steady control.

2. Sometimes liquids contain particles of insoluble solids that sink to the bottom of a test tube or beaker. Use one of the methods given below to separate a supernatant (the clear fluid) from insoluble solids.

  a. Figure B shows the proper method of decanting a supernatant liquid in a test tube.

  b. Figure C shows the proper method of decanting a supernatant liquid in a beaker by using a stirring rod. The rod should touch the wall of the receiving container. Hold the stirring rod against the lip of the beaker containing the supernatant liquid. As you pour, the liquid will run down the rod and fall into the beaker resting below. Using this method will prevent the liquid from running down the side of the beaker you are pouring from.

HEATING SUBSTANCES AND EVAPORATING SOLUTIONS

1. Use care in selecting glassware for high-temperature heating. The glassware should be heat resistant.

2. When using a gas flame to heat glassware, use a ceramic-centered wire gauze to protect glassware from direct contact with the flame. Wire gauzes can withstand extremely high temperatures and will help prevent glassware from breaking. **Figure D** shows the proper setup for evaporating a solution over a water bath.

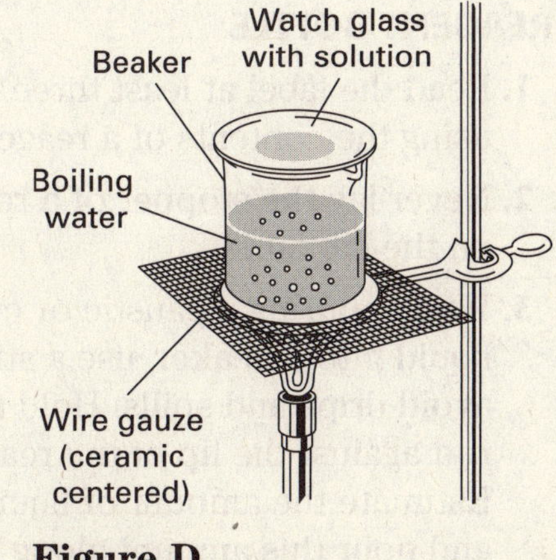

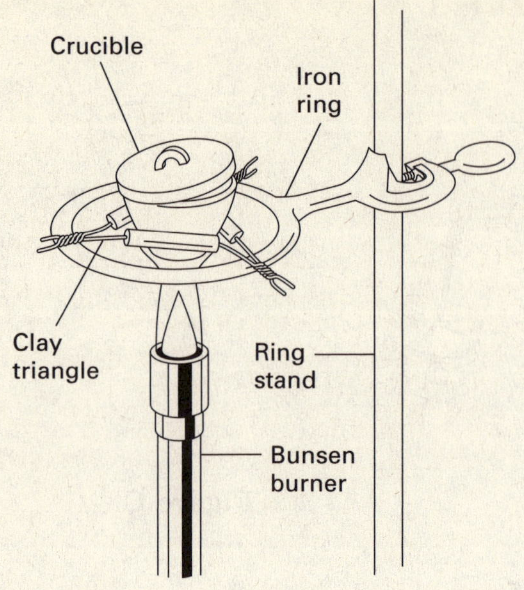

Figure E

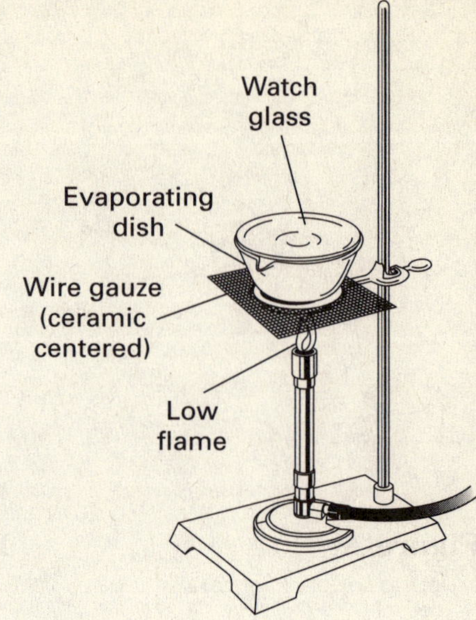

Figure F

3. In some experiments you are required to heat a substance to high temperatures in a porcelain crucible. **Figure E** shows the proper apparatus setup used to accomplish this task.

4. Figure F shows the proper setup for evaporating a solution in a porcelain evaporating dish with a watch glass cover that prevents spattering.

5. Glassware, porcelain, and iron rings that have been heated may look cool after they are removed from a heat source, but they can burn your skin even after several minutes of cooling. Use tongs, test tube holders, or heat-resistant mitts and pads whenever you handle this apparatus.

6. You can test the temperature of questionable beakers, ring stands, wire gauzes, or other pieces of apparatus that have been heated, by holding the back of your hand close to their surfaces before grasping them. You will be able to feel any heat generated from the hot surfaces. **Do not touch the apparatus until it is cool.**

POURING LIQUID FROM A REAGENT BOTTLE

1. Read the label at least three times before using the contents of a reagent bottle.

2. Never lay the stopper of a reagent bottle on the lab table.

3. When pouring a caustic or corrosive liquid into a beaker, use a stirring rod to avoid drips and spills. Hold the stirring rod against the lip of the reagent bottle. Estimate the amount of liquid you need, and pour this amount along the rod into the beaker. See **Figure G.**

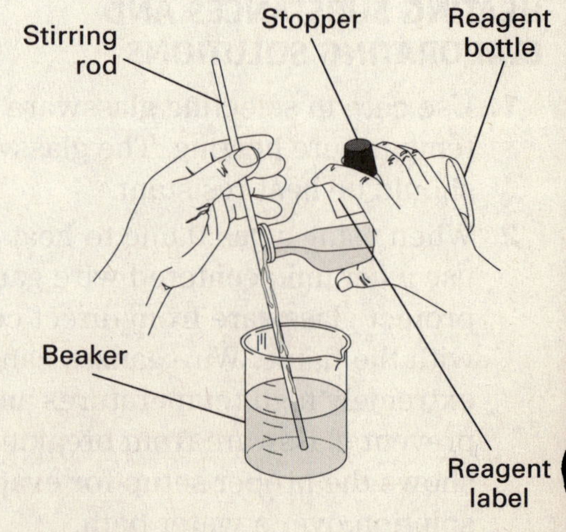

Figure G

4. Take extra precautions when handling a bottle of acid or strong base. Remember the following important rules: Never add water to any concentrated acid, particularly sulfuric acid, because the mixture can splash and will generate a lot of heat. To dilute any acid, add the acid to water in small quantities, while stirring slowly. Remember the "triple A's"—Always Add Acid to water.

5. Examine the outside of the reagent bottle for any liquid that has dripped down the bottle or spilled on the counter top. Your teacher will show you the proper procedures for cleaning up a chemical spill.

6. Never pour reagents back into stock bottles. At the end of the experiment, your teacher will tell you how to dispose of any excess chemicals.

HEATING MATERIAL IN A TEST TUBE

1. Check to see that the test tube is heat resistant.

2. Always use a test-tube holder or clamp when heating a test tube.

3. Never point a heated test tube at anyone, because the liquid may splash out of the test tube.

4. Never look down into the test tube while heating it.

5. Heat the test tube from the upper portions of the tube downward and continuously move the test tube, as indicated in **Figure H.** Do not heat any one spot on the test tube. Otherwise, a pressure buildup may cause the bottom of the tube to blow out.

USING A MORTAR AND PESTLE

1. A mortar and pestle should be used for grinding only one substance at a time. See **Figure I.**

2. Never use a mortar and pestle for simultaneously mixing different substances.

3. Place the substance to be broken up into the mortar.

4. Firmly push on the pestle to crush the substance. Then grind it to pulverize it.

5. Remove the powdered substance with a porcelain spoon.

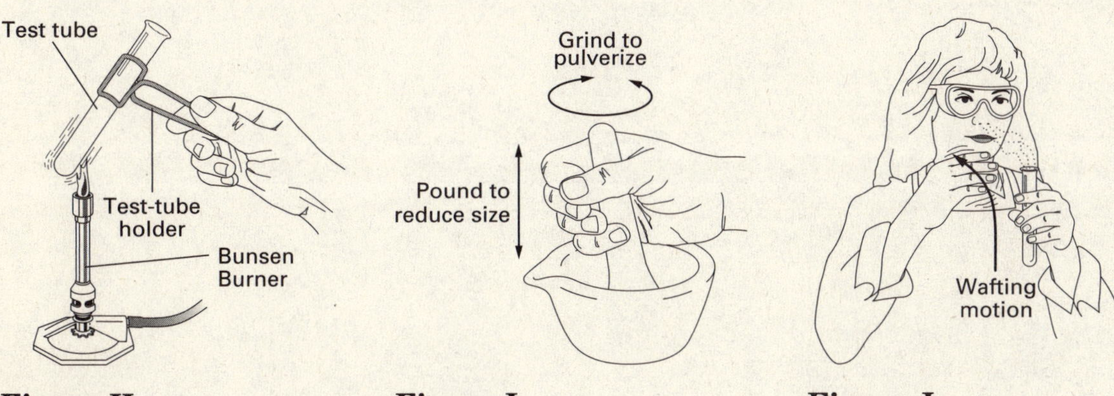

Figure H   **Figure I**   **Figure J**

DETECTING ODORS SAFELY

1. Test for the odor of gases by wafting your hand over the test tube and cautiously sniffing the fumes, as indicated in **Figure J.**

2. Do not inhale any fumes directly.

3. Use a fume hood whenever poisonous or irritating fumes are involved. **Do not** waft and sniff poisonous or irritating fumes.

Reactivity of Halide Ions

The four halide salts used in this experiment are found in your body. Although sodium fluoride is poisonous, trace amounts seem to be beneficial to humans in the prevention of tooth decay. Sodium chloride is added to most of our food to increase flavor while masking sourness and bitterness. Sodium chloride is essential for many life processes, but excessive intake appears to be linked to high blood pressure. Sodium bromide is distributed throughout body tissues, and in the past it has been used as a sedative. Sodium iodide is necessary for the proper operation of the thyroid gland, which controls cell growth. The concentration of sodium iodide is almost 20 times greater in the thyroid than in blood. The need for this halide salt is the reason that about 10 ppm of NaI is added to packages of table salt labeled "iodized."

The principal oxidation number of the halogens is -1. However, all halogens except fluorine may have other oxidation numbers. The specific tests you will develop in this experiment involve the production of recognizable precipitates and complex ions. You will use your observations to determine the halide ion present in an unknown solution.

MATERIALS

- 24-well microplate
- $AgNO_3$, 0.1 M
- $Ca(NO_3)_2$, 0.5 M
- gloves
- KBr, 0.2 M
- KI, 0.2 M
- lab apron
- $Na_2S_2O_3$, 0.2 M

- NaCl, 0.1 M
- NaF, 0.1 M
- NaOCl (commercial bleach), 5%
- $NH_3(aq)$, 4 M
- safety goggles
- starch solution, 3%
- thin-stemmed pipets (12)

Always wear safety goggles and a lab apron to protect your eyes and clothing. If you get a chemical in your eyes, immediately flush the chemical out at the eyewash station while calling to your teacher. Know the location of the emergency lab shower and eyewash station and the procedures for using them.

Do not touch any chemicals. If you get a chemical on your skin or clothing, wash the chemical off at the sink while calling to your teacher. Make sure you carefully read the labels and follow the precautions on all containers of chemicals that you use. If there are no precautions stated on the label, ask your teacher what precautions to follow. Do not taste any chemicals or items used in the laboratory. Never return leftovers to their original container; take only small amounts to avoid wasting supplies.

Call your teacher in the event of a spill. Spills should be cleaned up promptly, according to your teacher's directions.

Acids and bases are corrosive. If an acid or base spills onto your skin or clothing, wash the area immediately with running water. Call your teacher in the event of an acid spill. Acid or base spills should be cleaned up promptly.

Never put broken glass in a regular waste container. Broken glass should be disposed of separately according to your teacher's instructions.

Never stir with a thermometer because the glass around the bulb is fragile and might break.

OBJECTIVES

Observe the reactions of the halide ions with different reagents.

Analyze data to determine characteristic reactions of each halide ion.

Infer the identity of unknown solutions.

Procedure

1. Put on safety goggles, gloves, and a lab apron.

2. Put 5 drops of 0.1 M NaF into each of four wells in row A, as shown in **Figure 1.** Put 5 drops of 0.1 M NaCl into each of the wells in row B. Put 5 drops of 0.2 M KBr into each of the wells in row C and 5 drops of 0.2 M KI into each of the wells in Row D. Reserve rows E and F for unknown solutions.

Figure 1

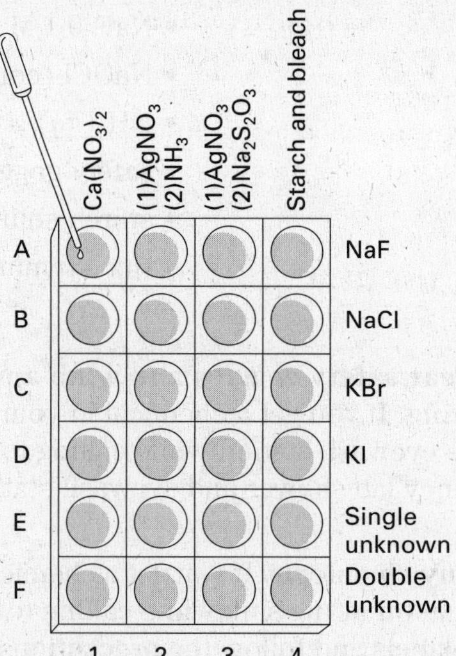

Reactivity of Halide Ions *continued*

3. Add 5 drops of 0.5 M $Ca(NO_3)_2$ solution to each of the four halide solutions in column 1. Record your observations in the **Table 1.**

4. Add 2 drops of 0.1 M $AgNO_3$ solution to each of the halides in columns 2 and 3. Record in **Table 1** the colors of the precipitates formed.

5. Add 5 drops of 4 M $NH_3(aq)$ to the precipitates in column 2. Record your observations in the **Table 1.**

6. Add 5 drops of 0.2 M $Na_2S_2O_3$ solution to the precipitates in column 3. Record your observations in **Table 1.**

7. To the halides in column 4, add 5 drops of starch solution and 1 drop of 5% bleach solution. Record your observations. Save the results of testing the known halide solutions for comparison with the tests of the unknown solutions.

8. Obtain an unknown solution. Put 5 drops of the unknown in each of the four wells in row E. Add the reagents to each well as you did in steps 3–6. Compare the results with those of the known halides in rows A–D. Record your findings in **Table 1,** and identify the unknown.

9. Obtain an unknown solution containing a mixture of two halide ions. Place 5 drops of the unknown mixture in each of the four wells in row F. Add the reagents to each well as you did in steps 3–6. Record your results. Compare the results with those of the known halides in rows A–D. Identify the halides in the double unknown solution.

10. Rinse the microplate into a trough or dishpan provided by your teacher. Clean all apparatus and your lab station. Return equipment to its proper place. Dispose of chemicals and solutions in the containers designated by your teacher. Do not pour any chemicals down the drain or in the trash unless your teacher directs you to do so. Wash your hands thoroughly before you leave the lab and after all work is finished.

TABLE 1: RESULTS OF THE REACTIONS OF HALIDE SALTS

Halide salts	$Ca(NO_3)_2$	$AgNO_3$	$AgNO_3$ + NH_3	$AgNO_3$ + $Na_2S_2O_3$	NaOCl + starch
NaF					
NaCl					
KBr					
KI					
Single unknown					
Double unknown					

Reactivity of Halide Ions *continued*

Analysis

1. Analyzing Data Which procedure(s) confirm(s) the presence of (a) F^- ions, (b) Cl^- ions, (c) Br^- ions, (d) I^- ions?

Conclusions

1. Drawing Conclusions What generalizations can be made about silver halides?

____

____

2. Applying Conclusions In nuclear explosions or accidents, iodine-131, a radioactive fission product, can become dispersed in the atmosphere. Eventually, the iodine isotope will fall onto the ground and be absorbed by plants. Explain how radiation from iodine-131 could become concentrated in the human body and cause a growth disorder.

____

____

____

3. Defending Conclusions Identify your unknown(s) and use your experimental evidence to support your identifications.

____

____

____

____

Periodicity of Properties of Oxides

Some oxides produce acidic solutions when they dissolve in water. These oxides are classified as acidic oxides (acid anhydrides), and they are the primary cause of acid rain. Other oxides produce basic solutions when they dissolve in water. These oxides are known as basic oxides (basic anhydrides). In general, nonmetal oxides produce acidic solutions and metallic oxides produce basic solutions.

To classify oxides as acidic or basic, you must test the solubility of the oxide in both a strong base and a strong acid. If the oxide is more soluble in a strong base, such as sodium hydroxide, the oxide is classified as an acidic oxide. If the oxide is more soluble in a strong acid, such as hydrochloric acid, the oxide is classified as a basic oxide. Those oxides that dissolve in both an acid and a base are called amphoteric oxides.

In this experiment you will investigate the properties of a number of oxides. From your observations and the relative positions of the oxides in the periodic table, you can generalize group and period trends in these properties.

OBJECTIVES

- **Determine** the solubility in water of several oxides.

- **Classify** these oxides as acidic or basic.

- **Compare** the solubility of two oxides in strong acid and in strong base.

- **Classify** these oxides as acidic, basic, or amphoteric.

- **Relate** experimental results to the positions of the elements in the periodic table, and generalize trends.

MATERIALS

- 6 M HCl
- 6 M NaOH
- carbonated beverage
- MgO
- SnO_2
- $SrCO_3$
- universal indicator (or pH paper)
- ZnO
- 24-well microplate
- 50 mL beaker
- 250 mL gas-collecting bottle
- Bunsen burner
- clay triangle
- crucible
- deflagrating spoon
- deionized water
- hot plate
- index card, 3×5 in.
- iron ring
- ring stand
- rubber stopper for bottle
- spatula
- stirring rod
- sulfur
- thermometer, nonmercury
- thin-stemmed pipet
- thin-stemmed pipet with two right-angle bends

Periodicity of Properties of Oxides *continued*

 **Always wear safety goggles and a lab apron, to protect your eyes and clothing.** If you get a chemical in your eyes, immediately flush the chemical out at the eyewash station while calling to your teacher. Know the locations of the emergency lab shower and eyewash station and the procedures for using them.

Do not touch any chemicals. If you get a chemical on your skin or clothing, wash the chemical off at the sink while calling to your teacher. Make sure you carefully read the labels and follow the precautions on all containers of chemicals that you use. If there are no precautions stated on the label, ask your teacher what precautions you should follow. Do not taste any chemicals or items used in the laboratory. Never return leftovers to their original containers; take only small amounts to avoid wasting supplies.

When using a Bunsen burner, confine long hair and loose clothing. Do not heat glassware that is broken, chipped, or cracked. Use tongs or a hot mitt to handle heated glassware and other equipment; heated glassware does not look hot. If your clothing catches fire, WALK to the emergency lab shower, and use it to put out the fire.

Call your teacher in the event of a spill. Spills should be cleaned up promptly, according to your teacher's directions.

Never put broken glass in a regular waste container. Broken glass and ceramic should be disposed of separately. Put the glass and ceramic pieces in the container designated by your teacher.

PROCEDURE

1. Place a small sample (about the size of an apple seed) of magnesium oxide, MgO, into a well of the microplate. Using a thin-stemmed pipet, add deionized water until the well is half full. Stir the solution with a stirring rod. Add three drops of universal indicator solution. Record the result in **Data Table 1.**

2. Add strontium carbonate, $SrCO_3$, to a crucible until it's one-third full. Set up a ring stand with an iron ring and clay triangle. Use a Bunsen burner with the ring stand to heat the strontium carbonate in the crucible until all the CO_2 has been driven off, approximately 10 min. Transfer some of the residue (SrO) to a well of the microplate, and repeat step **1** with SrO in place of MgO.

3. On a hot plate, heat 25 mL of water in a 50 mL beaker.

4. Using a thin-stemmed pipet, add deionized water to one well of the microplate until it is half full, and add three drops of universal indicator solution.

5. Obtain a thin-stemmed pipet with two right-angle bends. Fill this pipet half full with a carbonated beverage, and make sure that no beverage remains in the stem. Rinse off the outside of the pipet with deionized water.

Periodicity of Properties of Oxides *continued*

6. When the water in the beaker is between 70°C and 80°C, remove the beaker from the hot plate. Place the open end of the thin-stemmed pipet into the universal indicator solution in the well, and set the bulb into the beaker of hot water, as shown in **Figure A.** Allow the CO_2 to bubble for approximately 30 s, and record the results in **Data Table 1.**

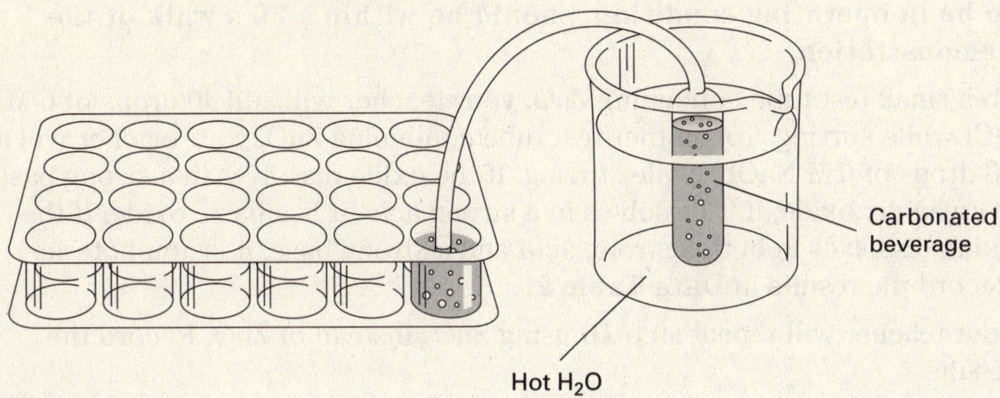

Figure A

7. Fill a 50 mL beaker half full with tap water.

CAUTION Because SO_2 is a very irritating gas, it should be prepared in a well-ventilated hood.

8. While working under a hood, pour 10 mL of deionized water into a gas-collecting bottle and add three drops of universal indicator solution. Fill a deflagrating spoon half full with sulfur, and insert the handle through the center of an index card. Use a Bunsen burner to heat the spoon until the sulfur ignites. Quickly move the spoon into the center of the gas-collecting bottle, as shown in **Figure B,** and slide the index card down until it covers the opening of the bottle.

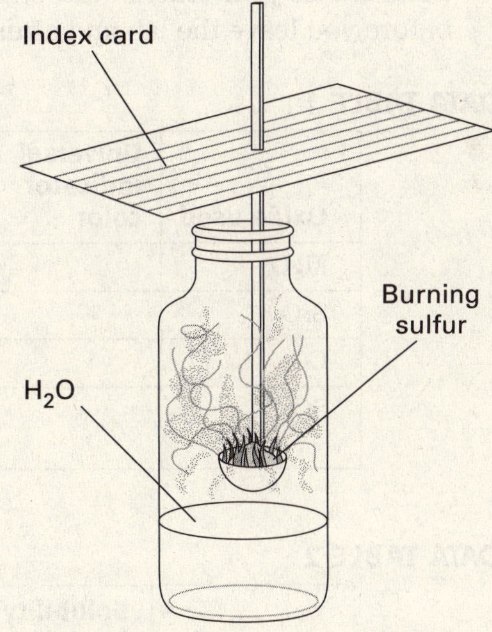

Figure B

9. After the sulfur has burned for 30 s, remove the spoon and put it into the 50 mL beaker of water to extinguish the sulfur. Stopper the bottle and shake it to dissolve the SO_2. Use this solution to fill a well of the microplate half full. Record your results in **Data Table 1.**

Periodicity of Properties of Oxides *continued*

Your teacher will perform steps **10** and **11**.
CAUTION The teacher will wear safety goggles, a face shield, a lab apron, and gloves while performing this demonstration. Students should wear safety goggles and lab aprons and stand behind a safety shield while observing. A safety shower and eyewash station, known to be in operating condition, should be within a 30 s walk of the demonstration.

10. To a small test tube containing ZnO, your teacher will add 40 drops of 6 M HCl while stirring. To another test tube containing ZnO, your teacher will add 40 drops of 6 M NaOH while stirring. If the oxide dissolves in a strong base, it is an acidic oxide; if it dissolves in a strong acid, it is a basic oxide. If the oxide dissolves in both a strong acid and a strong base, it is amphoteric. Record the results in **Data Table 2.**

11. Your teacher will repeat step **10** using SnO_2 instead of ZnO. Record the results.

DISPOSAL

12. Clean all apparatus and your lab station. Return equipment to its proper place. Dispose of chemicals and solutions in the containers designated by your teacher. Do not pour any chemicals down the drain or into the trash unless your teacher directs you to do so. Wash your hands thoroughly before you leave the lab and after all work is finished.

DATA TABLE 1

Oxide used	Universal indicator color	pH	Nature of oxide (acidic, basic, or amphoteric)
MgO			
SrO			
CO_2			
SO_2			

DATA TABLE 2

Oxide used	Solubility in 6 M HCl	Solubility in 6 M NaOH	Nature of oxide (acidic, basic, or amphoteric)
ZnO			
SnO_2			

Periodicity of Properties of Oxides *continued*

Analysis

1. **Organizing Results** For the oxides that you or your teacher tested, make a horizontal list of the group numbers of the elements forming the oxides, starting with Group 2. Write the formulas of the tested compounds under their group numbers. Under each formula, write whether you found the oxide to be acidic, basic, or amphoteric.

____

____

____

____

____

2. **Inferring Conclusions** What is the *general* trend in the properties of the oxides from left to right across the periodic table?

____

____

____

3. **Analyzing Results** For the two elements you tested in Group 14, what is the trend in properties within the group?

____

____

____

____

4. **Applying Ideas** Explain your answer to item **3** by classifying the two elements as metals or nonmetals.

____

____

____

____

____

____

5. **Predicting Outcomes** From what you have learned in this experiment, predict whether NO_2 would be found to be an acidic or a basic oxide. Explain your answer.

____

____

____

____

6. **Predicting Outcomes** From what you have learned in this experiment, predict whether SnO or PbO would be found to be the stronger basic anhydride. Explain your answer.

____

____

____

____

Periodicity of Properties of Oxides *continued*

Conclusions

1. **Applying Conclusions** Two test tubes containing insoluble oxides need to be cleaned. One contains a basic oxide, CaO, and the other contains an amphoteric oxide, As_4O_{10}. Suggest a way of cleaning these test tubes.

____

____

____

____

2. **Applying Conclusions** Lime, CaO, is often added to the lawns and gardens in regions where soils tend to be too acidic for some plants. Explain how this can help.

____

____

____

____

Conductivity as an Indicator of Bond Type

For a substance to conduct an electric current, the substance must possess free-moving charged particles. These charged particles may be delocalized electrons, such as those found in substances that form metallic bonds. The particles may also be mobile ions, such as those found in dissolved (or molten) salts, or they may be ions formed by certain molecular substances having polar covalent bonds that dissociate when dissolved in water.

Solid ionic substances that do *not* dissolve in water are usually considered to be *insulators*, substances that do not conduct electric current. Substances with nonpolar covalent bonds are also insulators, as are substances with polar covalent bonds that are not easily broken to form ions when they are dissolved in water.

The fact that a substance is a conductor in its pure form, or conducts only in solution, or is an insulator defines a physical property of the substance and provides clues about its inner structure and the type of bonding found in the substance. In this experiment, you will use qualitative measures of conductivity to determine the identities of eight unknown substances, identified as A, B, C, D, E, F, G, and H. Each letter corresponds to one of these substances: sodium sulfate, sugar, tap water, deionized (or distilled) water, tin, graphite, yellow chalk dust, and charcoal. These substances are not necessarily listed in the correct order. You will use your conductivity-test results and other information to identify each unknown as one of the substances listed.

OBJECTIVES

Observe and compare conductivities of various substances.

Relate conductivity to the type of bonds in a substance.

Infer the identities of unknown substances.

MATERIALS

- dropper bottle of distilled water
- glass stirring rod, small, or toothpicks
- gloves
- hand-held conductivity tester
- lab apron

- labeled containers 8, each with a different unknown substance
- paper towels
- safety goggles
- spatula
- spot plate
- thin-stemmed pipet

Always wear safety goggles, gloves, and a lab apron to protect your eyes and clothing. If you get a chemical in your eyes, immediately flush the chemical out at the eyewash station while calling to your teacher. Know the location of the emergency lab shower and eyewash station and the procedures for using them.

| Conductivity as an Indicator of Bond Type *continued*

 **Do not touch any chemicals.** If you get a chemical on your skin or clothing, wash the chemical off at the sink while calling to your teacher. Make sure you carefully read the labels and follow the precautions on all containers of chemicals that you use. If there are no precautions stated on the label, ask your teacher what precautions to follow. Do not taste any chemicals or items used in the laboratory. Never return leftovers to their original container; take only small amounts to avoid wasting supplies.

Never put broken glass in a regular waste container. Broken glass should be disposed of separately according to your teacher's instructions.

Procedure

1. Put on safety goggles, gloves, and a lab apron.

2. Examine each of the eight unknowns and record in your lab notebook your observations about their state (solid or liquid), color, texture, and other qualities.

3. Test your conductivity apparatus to make sure that it lights up when the two wires or leads touch different parts of a conducting substance. Your teacher will tell you what conducting substance to use.

4. Using a clean spatula for each solid and a clean pipet for each liquid, deliver a small sample of each unknown into a separate well of a spot plate. **Keep track of which unknown is in which well. The spatula should be wiped with a clean paper towel between each use to prevent contamination. Similarly, the pipet should be wiped with a clean paper towel and rinsed with distilled water between each use.** Put any water used for rinsing into the sink.

5. Place the wire leads in good contact with the first unknown sample, as shown in **Figure 1**. Note whether or not it conducts, as indicated by the light on the conductivity tester. (It may help to move the tester around a bit while watching the light.) If the two leads on the tester touch each other, the light will automatically go on. Do not mistake this for a positive test. Some results may be faint and require careful observations. Most pure substances are nonconductors, so don't expect a majority of your substances to conduct. Record all observations in your lab notebook.

Figure 1

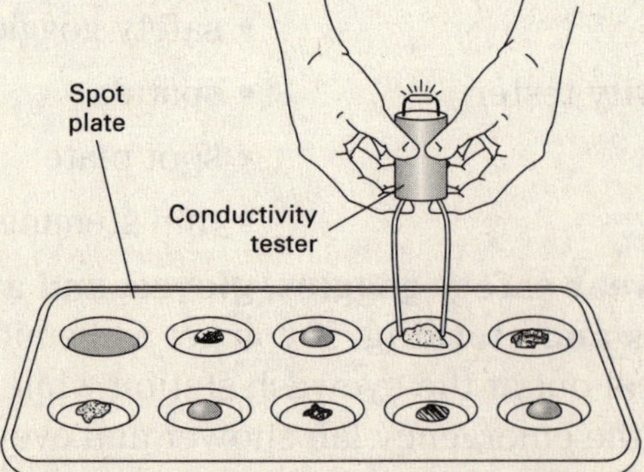

Conductivity as an Indicator of Bond Type *continued*

6. Repeat step **5** for the other unknown samples. **Be sure to thoroughly wipe the leads of the conductivity probe with a clean paper towel between each test.** For those substances that do conduct, compare the conductivities. Rate them on a scale from 0 to 4, with 4 as the best conductor. Record your observations and ratings in your lab notebook.

7. Using the pipet, add 10 drops of distilled water to each of the eight unknowns. **Hold the tip of the pipet 1 to 2 cm above the unknowns.** Use a small glass stirring rod or toothpicks to stir each mixture. **Be sure to wipe the stirring rod or toothpick thoroughly with a clean paper towel after each use.**

8. Observe each of the wells, and record in your lab notebook any observations about the dissolving or mixing of the unknowns.

9. Test the conductivity of each mixture, again wiping the leads thoroughly between each test. Record your results.

10. Clean all apparatus and your lab station. Return equipment to its proper place. Dispose of chemicals and solutions in the containers designated by your teacher. Do not pour any chemicals down the drain or into the trash unless your teacher directs you to do so. Wash your hands thoroughly before you leave the lab and after all work is finished.

Analysis

1. Organizing Ideas Which types of bonding are likely to be involved in the unknowns that were conductors in their pure form? Explain what free-moving charged particles are available in these types of bonds.

____

____

____

____

____

2. Organizing Ideas Which types of bonding are likely be involved in the unknowns that were not conductors in their pure form but conducted a current when they were mixed with water? Explain what free-moving charged particles are available in these types of bonds.

____

____

____

____

Conductivity as an Indicator of Bond Type *continued*

3. Organizing ideas Which types of bonding are likely to be involved in the unknowns that were not conductors in their pure form or when they were mixed with water? Explain why there were no free-moving charged particles available in these types of bonds.

____

____

____

____

4. Organizing Data and Relating Ideas Fill in the table below by using the Useful Information that follows to determine which type of bonding is likely to be present in each of the substances listed and whether the substances are likely to conduct an electric current. Predict which will be the best conductor in pure form and dissolved in water, which will conduct only marginally, and which will not conduct at all.

TABLE 1 CONDUCTIVITY OF VARIOUS SUBSTANCES

Substance	Type of bonding	Conductivity	Conductivity in H_2O	Matches unknown
Chalk				
Charcoal				
Graphite				
Sodium sulfate				
Sugar				
Tin				
Distilled water				
Tap water				

Conductivity as an Indicator of Bond Type *continued*

Useful Information

- Chalk is mostly $CaCO_3$ and dissolves only slightly in water.
- Sodium sulfate is white and dissolves in water.
- Tin is a metal.
- Charcoal is a form of carbon with the atoms bonded to each other in a randomly organized way.
- Graphite is a form of carbon. The atoms are bonded to each other in systematic sheets with some free electrons holding the sheets together.
- Sugar is a molecular substance that dissolves in water.
- Distilled or deionized water can be considered pure H_2O.
- Tap water contains many dissolved impurities that are molecular and ionic.

Conclusions

1. In the fifth column of your table, write the letter of the unknown that corresponds to the substance in each row. Explain your reasons for making each identification.

____

____

____

____

____

____

____

____

____

____

____

Conductivity as an Indicator of Bond Type *continued*

2. **Designing Experiments and Predicting Outcomes** Predict the conductivities of each of the materials listed below, and give reasons for your predictions. How would the conductivity of each material be affected if water were added to it? If your teacher approves, test your predictions for all the materials except liquid bromine, which is too dangerous to work with.

   a. an iron nail

   b. liquid bromine, Br_2

   c. wax (a random arrangement of molecules with long, nonpolar chains of carbon and hydrogen atoms)

   d. a dilute hydrochloric acid solution, $HCl(aq)$

   e. table salt, NaCl

____

____

____

____

____

____

____

____

Microscale

Chemical Bonds

Chemical compounds are combinations of atoms held together by chemical bonds. These chemical bonds are of two basic types—ionic and covalent. Ionic bonds result when one or more electrons from one atom or group of atoms is transferred to another atom. Positive and negative ions are created through the transfer. In covalent compounds no electrons are transferred; instead electrons are shared by the bonded atoms.

The physical properties of a substance, such as melting point, solubility, and conductivity, can be used to predict the type of bond that binds the atoms of the compound. In this experiment, you will test six compounds to determine these properties. Your compiled data will enable you to classify the substances as either ionic or covalent compounds.

OBJECTIVES

Compare the melting points of six solids.

Determine the solubilities of the solids in water and in ethanol.

Determine the conductivity of water solutions of the soluble solids.

Classify the compounds into groups of ionic and covalent compounds.

Summarize the properties of each group.

MATERIALS

- 24-well microplate
- calcium chloride
- candle
- citric acid
- conductivity tester
- ethanol
- gloves
- iron ring
- lab apron
- phenyl salicylate
- potassium iodide
- ring stand
- safety goggles
- sodium chloride
- sucrose
- tin can lid
- thin-stemmed pipets (2)

Always wear safety goggles, gloves, and a lab apron to protect your eyes and clothing. If you get a chemical in your eyes, immediately flush the chemical out at the eyewash station while calling to your teacher. Know the location of the emergency lab shower and eyewash station and the procedures for using them.

Do not touch any chemicals. If you get a chemical on your skin or clothing, wash the chemical off at the sink while calling to your teacher. Make sure you carefully read the labels and follow the precautions on all containers of chemicals that you use. If there are no precautions stated on the label, ask your teacher what precautions to follow. Do not taste any chemicals or items used in the laboratory. Never return leftovers to their original container; take only small amounts to avoid wasting supplies.

Do not heat glassware that is broken, chipped, or cracked. Use tongs or a hot mitt to handle heated glassware and other equipment because hot glassware does not always look hot.

When using a candle, confine long hair and loose clothing. If your clothing catches on fire, WALK to the emergency lab shower and use it to put out the fire.

Procedure

1. Put on safety goggles, gloves, and a lab apron.

2. Before you begin, write a brief description of each of the six substances in **Table 1.**

3. Place a can lid on an iron ring attached to a ring stand. Position the ring so that it is just above the tip of a candle flame, as shown in **Figure 1.** Light the candle for a moment to check that you have the correct height.

4. Place a few crystals of sucrose, sodium chloride, phenyl salicylate, calcium chloride, citric acid, and potassium iodide in separate locations on the lid, as shown in **Figure 2.** Do not allow the samples of crystals to touch. Draw and label a diagram that shows the position of each compound.

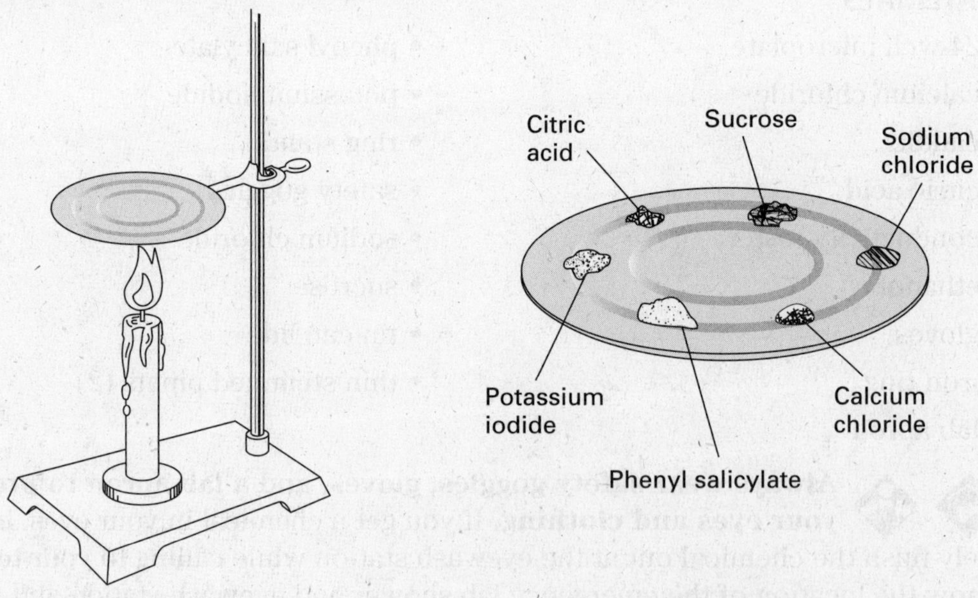

Figure 1   **Figure 2**

Chemical Bonds *continued*

5. For this experiment, it is not necessary to have exact values for the melting point. The lid will continue to get hotter as it is heated, so the order of melting will give relative melting points. Light the candle and observe. Note the substance that melts first by writing a *1* in **Table 1.** Record the order of melting for the other substances.

6. After 2 min, record an *n* in **Table 1** for each substance that did not melt. Extinguish the candle flame. Allow the can lid to cool while you complete the remainder of the experiment.

7. Put a *few* crystals of each of the white solids in the top row of your microtitration plate. Repeat with the second row. Add 10 drops of water to each well in the top row. Do not stir. Record the solubility of each substance in **Table 1.**

8. Add 10 drops of ethanol to each well in the second row of the microtitration plate. Do not stir. Record the solubility of each substance in **Table 1.**

9. Test the conductivity of each water solution in the top row by dipping both electrodes into each well of the microtitration plate. Be sure to rinse the electrodes and dry them with a paper towel after each test. If the bulb of the conductivity apparatus lights up, the solution conducts electric current. Record your results in **Table 1.**

10. Clean the microplate by rinsing it with water into a pan provided by your teacher. If any wells are difficult to clean, use a cotton swab. Wash your hands thoroughly before you leave the lab and after all work is finished.

TABLE 1 CHARACTERISTICS OF COMPOUNDS

Compound	Description	Melting point	Solubility in H_2O	Solubility in ethanol	Conductivity
Calcium chloride					
Citric acid					
Phenyl salicylate					
Potassium iodide					
Sodium chloride					
Sucrose					

Chemical Bonds *continued*

Analysis

1. Organizing Results Group the white substances into two groups according to their properties.

2. Organizing Results List the properties of each group.

Conclusions

1. Inferring Conclusions Use your textbook and your experimental data to determine which of the groups consists of ionic compounds and which consists of covalent compounds.

____

____

____

____

2. Relating Ideas Write a statement to summarize the properties of ionic compounds and another statement to summarize the properties of covalent compounds.

____

____

____

____

____

Name _________________________________ Class _______________ Date _____________

Tests for Iron(II) and Iron(III)

In this experiment, the complex hexacyanoferrate(II) ion (*ferro*cyanide), $Fe(CN)_6^{4-}$, and the hexacyanoferrate(III) ion (*ferri*cyanide), $Fe(CN)_6^{3-}$, will be used in identification tests for Fe^{2+} and Fe^{3+} ions. The charges on the two complex ions clearly indicate the difference in the oxidation state of the iron present in each. The (CN) group in each complex ion has a charge of -1. Thus, iron(II) is present in the *ferro*cyanide ion, $[Fe^{2+}(CN^-)_6]^{4-}$. Iron(III) is present in the *ferri*cyanide ion group, $[Fe^{3+}(CN^-)_6]^{3-}$. A deep blue precipitate results when either complex ion combines with iron in a different oxidation state from that present in the complex. The deep blue color of the precipitate is caused by the presence of iron in both oxidation states. The color provides a means of identifying either iron ion. If the deep blue precipitate is formed on addition of the $[Fe^{2+}(CN^-)_6]^{4-}$ complex, the iron ion responsible must be the iron(III) ion. Similarly, a deep blue precipitate formed with the $[Fe^{3+}(CN^-)_6]^{3-}$ complex indicates the presence of the iron(II) ion.

Both of the deep blue precipitates are known to have the same composition. The potassium salt of the complex ion has the formula $KFeFe(CN)_6 \cdot H_2O$.

The thiocyanate ion, SCN^-, provides a test for confirming the presence of Fe^{3+} ion. The soluble $FeSCN^{2+}$ complex imparts a blood red color to the solution.

OBJECTIVES

Observe tests of known solutions containing iron(II) or iron(III) ions.

Compare results for the two ions and infer conclusions.

Design a procedure for identifying the two ions in one solution.

MATERIALS

- $FeCl_3$, 0.1 M
- $Fe(NH_4)_2(SO_4)_2$, 0.2 M
- gloves
- $K_3Fe(CN)_6$, 0.1 M
- $K_4Fe(CN)_6$, 0.1 M
- KSCN, 0.2 M

- lab apron
- plastic wrap, 8 cm × 30 cm
- safety goggles
- sheet of paper, white
- thin-stemmed pipets (5)

Always wear safety goggles, gloves, and a lab apron to protect your eyes and clothing. If you get a chemical in your eyes, immediately flush the chemical out at the eyewash station while calling to your teacher. Know the location of the emergency lab shower and eyewash station and the procedures for using them.

Tests for Iron(II) and Iron(III) *continued*

 **Do not touch any chemicals.** If you get a chemical on your skin or clothing, wash the chemical off at the sink while calling to your teacher. Make sure you carefully read the labels and follow the precautions on all containers of chemicals that you use. If there are no precautions stated on the label, ask your teacher what precautions to follow. Do not taste any chemicals or items used in the laboratory. Never return leftover chemicals to their original containers; take only small amounts to avoid wasting supplies.

Call your teacher in the event of a spill. Spills should be cleaned up promptly, according to your teacher's directions.

Acids and bases are corrosive. If an acid or base spills onto your skin or clothing, wash the area immediately with running water. Call your teacher in the event of an acid spill. Acid or base spills should be cleaned up promptly.

Procedure

1. Put on safety goggles, gloves, and a lab apron.

2. Check that you have a labeled pipet for each solution listed in the materials.

3. Place the plastic-wrap rectangle on a white sheet of paper.

4. Along the top of the plastic wrap, place 5 drops of a freshly prepared iron(II) ammonium sulfate solution in each of the three locations shown in **Figure 1.**

5. Along the bottom of the plastic wrap, place 5 drops of a freshly prepared iron(III) chloride solution, as shown in **Figure 1.**

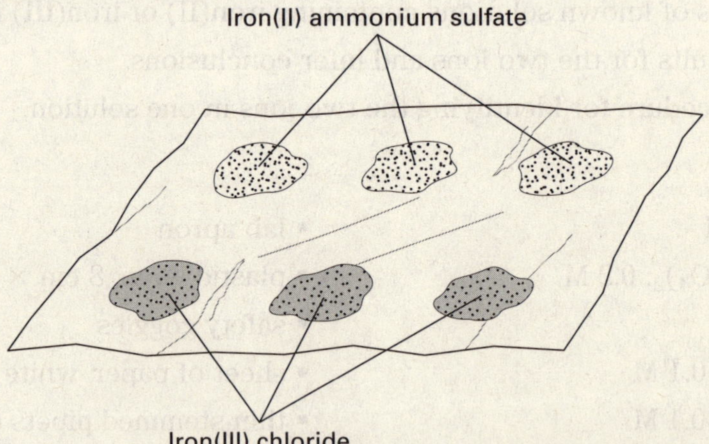

6. Add 1 drop of 0.1 M $K_4Fe(CN)_6$ solution to the first sample of iron(II) ions at the top of the plastic wrap and 1 drop to the first sample of iron(III) ions at the bottom. Record your observations in **Table 1.**

7. Add 1 drop of 0.1 M KSCN solution to the second sample of iron(II) ions and 1 drop to the second sample of iron(III) ions. Record your observations in **Table 1.**

8. Add 1 drop of 0.1 M $K_3Fe(CN)_6$ solution to the third sample of iron(II) ions and 1 drop to the third sample of iron(III) ions. Record your observations.

9. Clean all apparatus and your lab station. Return equipment to its proper place. Dispose of chemicals and solutions in the containers designated by your teacher. Do not pour any chemicals down the drain or in the trash unless your teacher directs you to do so. Wash your hands thoroughly before you leave the lab and after all work is finished.

TABLE 1: RESULTS OF TESTS FOR IRON(II) AND IRON(III)

Iron ion	Hexacyanoferrate(II) ion $[Fe^{2+}(CN^-)_6]^{4-}$	Hexacyanoferrate(III) ion $[Fe^{3+}(CN^-)_6]^{3-}$	Thiocyanate ion SCN^-
Fe^{2+}			
Fe^{3+}			
Observations in step 6			
Observations in step 7			
Observations in step 8			

Analysis

1. Organizing Ideas Explain specifically how you would make a conclusive test for an iron(III) salt.

____

____

____

____

2. Organizing Ideas Which test for iron(II) ions is conclusive?

____

____

3. Relating Ideas When iron(II) ammonium sulfate was mixed with the $[Fe^{2+}(CN^-)_6]^{4-}$ ion, the precipitate was initially white but turned blue upon exposure to air. What happened to the iron(II) ion when the precipitate turned blue?

____

____

Tests for Iron(II) and Iron(III) *continued*

Conclusions

1. **Designing Experiments** Suppose you have a solution containing both an iron(II) salt and an iron(III) salt. How would you proceed to identify both Fe^{2+} and Fe^{3+} in this solution?

____

____

____

2. **Relating Ideas** Blueprint paper can be made by soaking paper in a brown solution of $[Fe^{3+}(CN^-)_6]^{3-}$ and iron(III) ammonium citrate. Wherever the paper is exposed to bright light, the paper turns blue. Explain why this happens.

____

____

____

Simple Qualitative Analysis

If an unknown sample is one of a limited number of possible compounds, a simple test often can determine its identity. For example, a flame test can distinguish between KCl and $NaNO_3$. A drop of potassium hexacyanoferrate(III) solution can tell you whether $FeCl_2$ or $FeCl_3$ is present. There are hundreds of simple qualitative tests such as these to distinguish among a few possibilities.

Qualitative tests are important to the forensic chemist, one who is interested in solving crimes, but they have been largely replaced by instrumental analysis, which is fast and requires a very small sample. For example, instrumental analysis can detect as little as 1×10^{-11} g of 9-tetrahydrocannabinol (one of the active ingredients in marijuana) in 1 mL of blood plasma. Unfortunately, this kind of instrumentation (a gas chromatograph connected to a mass spectrograph) is expensive, so simple chemical tests are still often useful. A drop of hydrochloric acid is enough to allow the police department's forensic chemist to distinguish between a bag of cocaine and a bag of baking soda.

In this experiment you will identify the contents of a number of vials. The substance in each vial is one of the two compounds listed on the label. To decide which compound is present, you will make a few simple tests.

OBJECTIVES

Observe qualitative tests using known ionic compounds.

Decide which qualitative test to use in identifying an unknown ionic compound.

Describe the chemistry of common ionic compounds.

MATERIALS

- beaker, 50 mL
- Bunsen burner and related equipment
- cobalt glass plates (2)
- filter paper, cut into short strips
- flame-test wire
- gloves
- HCl, 1.0 M
- $KMnO_4$, 0.1 M
- lab apron
- Na_2CO_3
- NaOH, 1.0 M
- Na_2SO_3
- NH_4Cl
- red litmus paper
- safety goggles
- sparker
- spot plate or small test tubes (3)

| Simple Qualitative Analysis *continued*

Always wear safety goggles and a lab apron to protect your eyes and clothing. If you get a chemical in your eyes, immediately flush the chemical out at the eyewash station while calling to your teacher. Know the location of the emergency lab shower and eyewash station and the procedures for using them.

Do not touch any chemicals. If you get a chemical on your skin or clothing, wash the chemical off at the sink while calling to your teacher. Make sure you carefully read the labels and follow the precautions on all containers of chemicals that you use. If there are no precautions stated on the label, ask your teacher what precautions to follow. Do not taste any chemicals or items used in the laboratory. Never return leftovers to their original container; take only small amounts to avoid wasting supplies.

Call your teacher in the event of a spill. Spills should be cleaned up promptly, according to your teacher's directions.

Acids and bases are corrosive. If an acid or base spills onto your skin or clothing, wash the area immediately with running water. Call your teacher in the event of an acid spill. Acid or base spills should be cleaned up promptly.

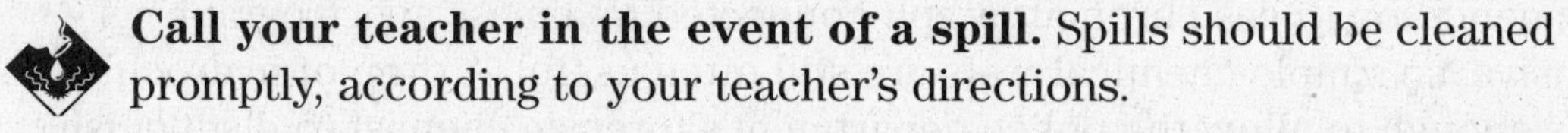

Never put broken glass in a regular waste container. Broken glass should be disposed of separately according to your teacher's instructions.

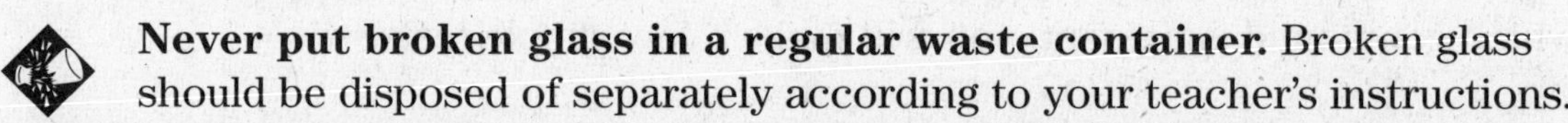

Do not heat glassware that is broken, chipped, or cracked. Use tongs or a hot mitt to handle heated glassware and other equipment because hot glassware does not always look hot.

When using a Bunsen burner, confine long hair and loose clothing. If your clothing catches on fire, WALK to the emergency lab shower and use it to put out the fire. Do not heat glassware that is broken, chipped, or cracked. Use tongs or a hot mitt to handle heated glassware and other equipment because hot glassware does not always look hot.

Procedure

1. Put on safety goggles, gloves, and a lab apron.

2. Moisten a strip of filter paper with 0.1 M $KMnO_4$. Place a few crystals of sodium sulfite on a spot plate or in a small test tube, and add 2 drops of 1.0 M HCl. Immediately hold the strip of filter paper over the crystals, as shown in **Figure 1.** This reaction is characteristic of the sulfite ion, SO_3^{2-}. Record your observations.

Simple Qualitative Analysis *continued*

Figure 1

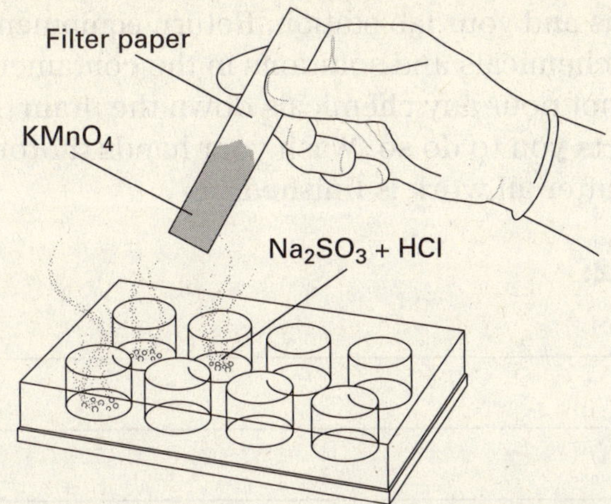

3. Place a few crystals of sodium carbonate on a spot plate or in a small test tube. Add 2 drops of 1.0 M HCl. Immediately hold a strip of filter paper moistened with 0.1 M KMnO$_4$ over the solution. This reaction is characteristic of the carbonate ion, CO_3^{2-}. Record your observations.

4. Place a few crystals of ammonium chloride on a spot plate or in a small test tube. Add 2 drops of 1.0 M NaOH. Quickly hold a piece of moistened red litmus paper over the solution, as shown in **Figure 2.** This reaction is characteristic of the ammonium ion, NH_4^+. Record your observations.

Figure 2

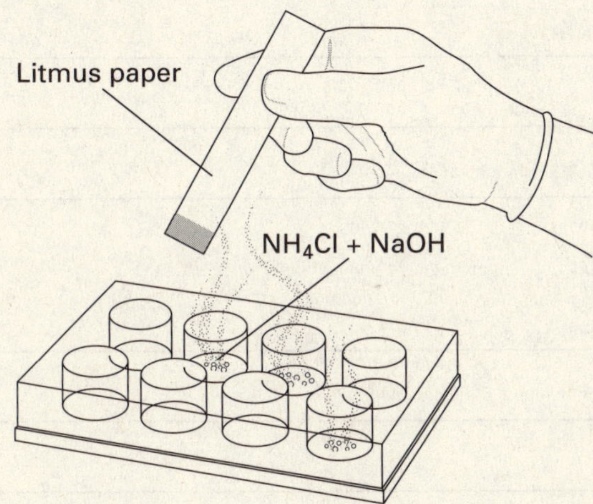

5. Label a 50 mL beaker *Waste* and add about 10 mL of HCl to the beaker. Clean the flame-test wire by dipping it in the HCl and then holding it in the colorless flame of the Bunsen burner. Put 10 drops of Na$_2$CO$_3$ into one of the wells. Dip the wire in the Na$_2$CO$_3$ solution, and then hold it in the Bunsen burner flame. Record the color of the flame.

6. You now have a number of simple tests to identify each of the unknowns listed in **Table 1.** If a solution is needed for a test, dissolve a small amount of the compound in water. Consider all four ions in the two compounds for each unknown. For example, unknown 1 contains K$^+$, NO$_3^-$, Na$^+$, and Cl$^-$ ions. You should be able to predict the results for each compound before you begin the test.

▍Simple Qualitative Analysis *continued*

7. Clean all apparatus and your lab station. Return equipment to its proper place. Dispose of chemicals and solutions in the containers designated by your teacher. Do not pour any chemicals down the drain or in the trash unless your teacher directs you to do so. Wash your hands thoroughly before you leave the lab and after all work is finished.

Observations, step 2:

____

____

____

Observations, step 3:

____

____

____

Observations, step 4:

____

____

____

Observations, step 5:

____

____

____

Simple Qualitative Analysis *continued*

TABLE 1: TEST RESULTS

Unknown	Test Used	Observations	Identity
KNO_3 or $NaCl$			
NH_4Cl or $MgCO_3$			
$LiNO_3$ or Na_2CO_3			
$Sr(NO_3)_2$ or Na_2SO_3			
$(NH_4)_2CO_3$ or $ZnSO_4$			
$BaCl_2$ or $NaNO_3$			
Na_2SO_3 or Na_2CO_3			

Analysis

1. **Relating Ideas** Explain how to generate ammonia gas in the laboratory. Write and balance the equation for the reaction.

   ____

   ____

   ____

2. **Relating Ideas** Give the formula and the name of a compound that gives a violet color to a flame and, when HCl is added, produces bubbles of gas that turn potassium permanganate brown.

   ____

   ____

Conclusions

1. **Analyzing Results** Which salt in the table will react with both HCl and NaOH? Write the equations for the reactions.

   ____

   ____

   ____

Generating and Collecting O$_2$

When oxygen reacts with another substance and large amounts of energy are generated as heat and light, the reaction is called burning or combustion. In these reactions, oxygen is referred to as supporting combustion, because oxygen itself is not considered a combustible gas. What is the nature of combustion? How does combustion in an environment of 100% oxygen compare with combustion in air which is about 20% oxygen? To answer these questions, you will generate and collect some oxygen gas and test it yourself to discover some of its chemical and physical properties.

OBJECTIVES

Observe the production of O$_2$.

Use the technique of water displacement to collect O$_2$ gas.

Relate chemical and physical properties to observations.

Apply conclusions to larger settings.

Formulate hypotheses.

Compare reactions of O$_2$ gas in different situations.

MATERIALS

- beaker, 100 mL or larger (2)
- Bunsen burner or candle
- forceps
- gloves
- H$_2$O$_2$ solution, 3%
- KI solution, 1.0 M
- lab apron
- micro O$_2$ generator
- paper towels
- pegs, small, to fit holes in stoppers (5)
- Petri dish, half
- rubber stoppers, one-holed (5)
- safety goggles
- steel wool
- tap water
- test tubes, small (7)
- wooden splints

Always wear safety goggles, gloves, and a lab apron to protect your eyes and clothing. If you get a chemical in your eyes, immediately flush the chemical out at the eyewash station while calling to your teacher. Know the location of the emergency lab shower and eyewash station and the procedures for using them.

Do not touch any chemicals. If you get a chemical on your skin or clothing, wash the chemical off at the sink while calling to your teacher. Make sure you carefully read the labels and follow the precautions on all containers of chemicals that you use. If there are no precautions stated on the label, ask your teacher what precautions to follow. Do not taste any chemicals or items used in the laboratory. Never return leftover chemicals to their original containers; take only small amounts to avoid wasting supplies.

| Generating and Collecting O₂ *continued*

Call your teacher in the event of a spill. Spills should be cleaned up promptly, according to your teacher's directions.

Do not heat glassware that is broken, chipped, or cracked. Use tongs or a hot mitt to handle heated glassware and other equipment because hot glassware does not always look hot.

When using a Bunsen burner, confine long hair and loose clothing. If your clothing catches on fire, WALK to the emergency lab shower and use it to put out the fire. A beaker of water should be kept near at all times, in case you need to quickly extinguish a burning wood splint.

Procedure

1. Put on safety goggles, gloves, and a lab apron.

2. Label one beaker *water* and the other *spent solution*. Fill the appropriately labeled beaker with water. Be sure both beakers are nearby as you proceed.

3. Label the test tubes *1* through *7*. Fill test tubes *2*, *4*, *5*, *6*, and *7* completely with water. Insert the one-holed stoppers without creating any air bubbles in the test tubes. (Leave test tubes *1* and *3* empty and open in the test-tube rack—they will serve as controls.)

4. The micro O₂ generator is a plastic vial capped with a lid having a nozzle, as shown in **Figure 1.**

Make sure your O₂ generator cap has a nozzle and that the nozzle is not plugged in any way before continuing the experiment.

Remove the cap, and carefully add enough 3% H₂O₂ to fill the vial to within 1 cm of the top. Then add 3 or 4 drops of KI solution. Replace the cap and set the O₂ generator in the Petri dish. Record your observations.

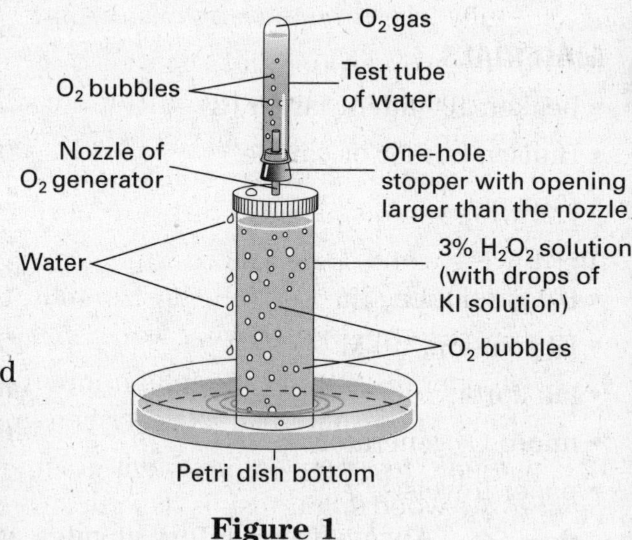

Figure 1

Observations:

____

____

____

Generating and Collecting O$_2$ *continued*

5. To collect the gases, place test tube *2* mouth downward over the nozzle of the generator, as shown in **Figure 1.** As soon as the test tube is completely filled with oxygen, remove it from the nozzle and plug the stopper hole with a small peg to prevent the collected gas from escaping. Repeat this procedure for test tubes *4, 5, 6,* and *7.*

6. If the reaction slows down so that it takes more than 1 min to collect a test tube of oxygen gas, uncap the vial, decant the liquid into the *spent solution* beaker, and replace it with fresh H$_2$O$_2$ solution and drops of KI solution. Then replace the cap and nozzle, and resume collecting gas.

For steps 7–13, read the directions carefully, and predict what you think might happen. Write your predictions in Table 1. Then follow the directions, and record your observations alongside your predictions.

7. Flaming wood splint and test tube *1* (control): Light the Bunsen burner or candle and use it to light one end of a wood splint. **Keep the gas-generating vial away from the flame and wood splint!** Hold test tube *1* (control test tube—filled with air) horizontally, and carefully insert the flaming wood splint. Record your observations in **Table 1.**

8. Flaming wood splint and test tube *2* (oxygen): Relight the wood splint, remove the stopper from test tube *2*, hold the test tube horizontally, and carefully insert the burning wood splint. Record your observations.

9. Glowing wood splint and test tube *3* (control): Light the end of the wood splint again. Let it burn for 5 to 10 seconds, and then blow out the flame, but make sure a glowing ember persists on the tip of the wood splint. Hold test tube *3* (control test tube—filled with air) horizontally, and carefully insert the glowing wood splint. Record your observations.

10. Glowing wood splint and test tube *4* (oxygen): Repeat **step 9** using test tube *4* and holding the test tube horizontally. Carefully unstopper the test tube before inserting the glowing wood splint. Record your observations.

11. Unstopper test tube *5* and hold it mouth upward for 2 minutes. Then try the glowing wood splint test as described in **step 9.** Record your observations.

12. Unstopper test tube *6* and hold it mouth downward for 2 minutes. Then try the glowing wood splint test as described in **step 9.** Record your observations.

13. Twist a few strands of steel wool into a single yarnlike string. Fold it in half, and hold the two loose ends together with forceps. Hold the folded bend in the Bunsen burner or candle flame for a few seconds, and then remove it and observe. Repeat this procedure, but this time quickly and carefully insert the glowing steel wool into test tube *7.* Record your observations.

14. If your teacher approves, refill the test tubes first with water and then with oxygen gas, as you did before, and recheck the observations you made.

| Generating and Collecting O$_2$ *continued*

15. Clean all apparatus and your lab station. Return equipment to its proper place. Dispose of chemicals and solutions in the containers designated by your teacher. Do not pour any chemicals down the drain or in the trash unless your teacher directs you to do so. Make sure to shut off the gas valve completely before leaving the laboratory. Wash your hands thoroughly before you leave the lab and after all work is finished.

TABLE 1 EXPERIMENTAL OBSERVATIONS

Test-tube number	Predictions	Observations
1. Flame, air		
2. Flame, O$_2$		
3. Ember, air		
4. Ember, O$_2$		
5. Mouth up, ember		
6. Mouth down, ember		
7. Lit steel wool, O$_2$		

Analysis

1. Analyzing Information What evidence was there that a reaction was occurring inside the generator?

____

2. Relating Ideas Write a balanced chemical equation for the O$_2$ generator reaction. (Hint: include KI in the equation, but only as a catalyst.)

3. Relating Ideas Write a word equation for the wood-splint tests using the following reactants and products: ash, wood, carbon dioxide, water vapor, and oxygen.

Generating and Collecting O$_2$ *continued*

Conclusions

1. Inferring Conclusions and Relating Ideas What caused the burning wood splint to eventually be extinguished in test tube *1*? (Hint: refer to the word equation you wrote in answer to Analysis question 3.)

____

____

2. Inferring Conclusions and Resolving Discrepancies Compare your results for test tubes *1* and *2*. Explain the differences you observed by considering which of the two reactants, wood or oxygen, is the limiting reactant for the reaction.

____

____

____

____

____

3. Inferring Conclusions and Resolving Discrepancies Compare your results for test tubes *3* and *4*. Why didn't the ember in test tube *3* behave like the ember in test tube *4*? (Hint: Which reaction was quicker and more thorough? Why?)

____

____

____

____

4. Inferring Conclusions and Resolving Discrepancies How does the "burning" of steel wool in air compare with the burning of steel wool in pure O$_2$ (test tube *7*)? Explain.

____

____

____

____

Generating and Collecting O_2 *continued*

5. Organizing Conclusions Using your observations from this experiment, list as many physical properties of oxygen as you can. Explain what evidence you have for each property. (At least four properties were observable in this experiment.)

____

____

____

____

____

____

6. Predicting Outcomes If you were to put a plain wood splint (with no ember or flame) into a tube full of oxygen, would it burst into flames? Would an ember form? Explain your answer. (Hint: what is required for a chemical reaction to occur in addition to having the reactant(s) present?)

____

____

____

____

Generating and Collecting H₂

Hydrogen gas has the lowest density of all the substances found on Earth. Once, hydrogen was used in blimps and dirigibles. Today, however, such aircraft use only helium, even though helium is less buoyant and considerably more difficult to obtain. Why was helium substituted for hydrogen? What was wrong with using hydrogen? To answer these questions, you will generate and collect small quantities of hydrogen gas and test the samples to find out for yourself some of the properties of hydrogen.

OBJECTIVES

Observe the production of H_2.

Use the technique of water displacement to collect H_2 gas.

Formulate hypotheses for reactions of H_2 gas in different situations.

MATERIALS

- beaker, 100 mL or larger (2)
- Bunsen burner or candle
- $CuSO_4$, 0.2 M
- forceps or tongs (optional)
- gloves
- HCl, 1.0 M
- lab apron
- micro H_2 generator
- pegs, small, to fit holes in stoppers (5)
- Petri dish, half
- rubber stoppers, one-holed (5)
- safety goggles
- tap water
- test tubes, small (5)
- wooden splints

Always wear safety goggles, gloves, and a lab apron to protect your eyes and clothing. If you get a chemical in your eyes, immediately flush the chemical out at the eyewash station while calling to your teacher. Know the location of the emergency lab shower and eyewash station and the procedures for using them.

Do not touch any chemicals. If you get a chemical on your skin or clothing, wash the chemical off at the sink while calling to your teacher. Make sure you carefully read the labels and follow the precautions on all containers of chemicals that you use. If there are no precautions stated on the label, ask your teacher what precautions to follow. Do not taste any chemicals or items used in the laboratory. Never return leftovers to their original container; take only small amounts to avoid wasting supplies.

Call your teacher in the event of a spill. Spills should be cleaned up promptly, according to your teacher's directions.

Acids and bases are corrosive. If an acid or base spills onto your skin or clothing, wash the area immediately with running water. Call your teacher in the event of an acid spill. Acid or base spills should be cleaned up promptly.

Do not heat glassware that is broken, chipped, or cracked. Use tongs or a hot mitt to handle heated glassware and other equipment because hot glassware does not always look hot.

When using a Bunsen burner, confine long hair and loose clothing. If your clothing catches on fire, WALK to the emergency lab shower and use it to put out the fire. A beaker of water should be kept near at all times, in case you need to quickly extinguish a burning wood splint.

Procedure

1. Put on safety goggles, gloves, and a lab apron.

2. Label one beaker *water* and the other *spent solution*. Fill the appropriately labeled beaker with water. Be sure both beakers are nearby as you proceed.

3. Label the test tubes *1, 2, 3, 4,* and *5*. Fill test tubes 1, 3, 4, and 5 completely full of water. Fill test tube 2 only half way. Insert the one-holed stoppers into all five test tubes. Avoid creating air bubbles in the test tubes.

4. The micro H₂ generator is a plastic vial containing several pieces of zinc metal capped with a lid that has a nozzle.

Make sure your H₂ generator cap has a nozzle and that the nozzle is not plugged before continuing with the experiment.

Remove the cap, and carefully add enough 1.0 M HCl to fill the vial to within 1 cm of the top. Replace the cap, and set the H₂ generator in the petri dish. Add a few drops of CuSO₄ solution as a catalyst. Observe the reaction, and record your observations.

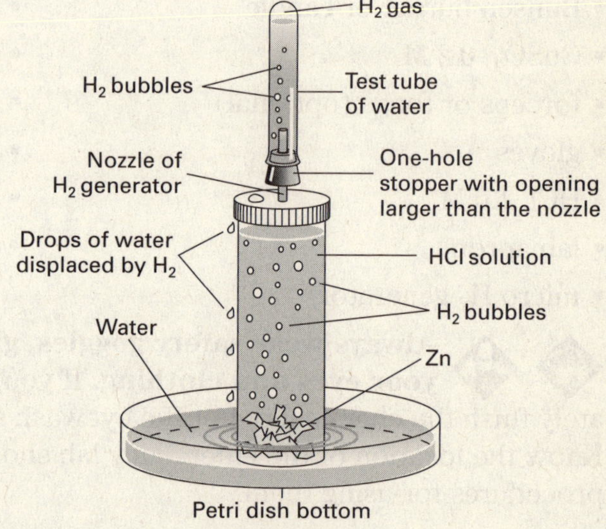

Figure 1

Observations:

____

____

____

Generating and Collecting H$_2$ *continued*

5. To collect the gas, place water-filled test tube 1 upside down over the nozzle of the generator, as shown in **Figure 1.** As soon as the test tube is completely filled with hydrogen, remove it from the nozzle and plug the stopper hole with a small peg to prevent the collected gas from escaping. Repeat this procedure using test tubes 2, 3, 4, and 5.

6. If the reaction slows down so that it takes more than 1 min to collect the tube of hydrogen gas, lift off the tube and uncap the vial. Decant the remaining liquid into the beaker labeled *spent solution* and replace it with fresh solution. Replace the cap and nozzle, and resume collecting gas.

For **steps 7** through **11,** read the directions carefully and predict what might happen. Write your predictions in **Table 1.** Then follow the directions, and record your observations alongside your predictions in the data table.

Because this investigation is carried out on a microscale level, the pop tests in steps 7 through 11, though potentially loud, are completely safe. The test tubes may be held with forceps or tongs or in your hand. Your teacher will tell you how you should hold the test tubes for the pop tests. Be sure to keep the gas-generating vial away from the flame and the wooden splint!

7. Flaming wooden splint and test tube 1 (hydrogen): Light the Bunsen burner or candle, and use it to light one end of a wood splint. Remove the stopper from test tube 1, and carefully insert the burning splint into the mouth of the test tube. Record your observations in **Table 1.**

8. Flaming wooden splint and test tube 2 (half hydrogen): Repeat the pop test described in **step 7** with test tube 2, which contains only half as much hydrogen as test tube 1 did. Record your observations in **Table 1.**

9. Glowing wooden splint and test tube 3 (hydrogen): Repeat the pop test with test tube 3, but this time blow out the flame, making sure a glowing ember persists on the tip of the wooden splint as you insert it into the test tube. Record your observations.

10. Unstopper test tube 4 and hold it mouth upward for 30 s; then try the pop test. Record your observations.

11. Unstopper test tube 5 and hold it mouth downward for 30 s; then try the pop test. Record your observations.

12. If your teacher approves, refill the test tubes first with water and then with hydrogen gas and recheck any of the observations you made. Record all observations.

13. Clean all apparatus and your lab station. Return equipment to its proper place. Dispose of chemicals and solutions in the containers designated by your teacher. Do not pour any chemicals down the drain or in the trash unless your teacher directs you to do so. Shut off the gas valve completely before leaving the laboratory. Wash your hands thoroughly before you leave the lab.

Generating and Collecting H$_2$ *continued*

TABLE 1 EXPERIMENTAL OBSERVATIONS

Test-tube number	Predictions	Observations
1. Flame, pure H$_2$		
2. Flame, 50% H$_2$		
3. Ember, pure H$_2$		
4. Mouth up, flame		
5. Mouth down, flame		

Analysis

1. Analyzing Information What evidence was there that a reaction was occurring inside the generator?

____

____

2. Relating Ideas Write a balanced chemical equation for the H$_2$ generator reaction.

3. Relating Ideas Write a balanced chemical equation for the reaction taking place during the pop test. (Hint: Water vapor is the product of the reaction.)

| Generating and Collecting H$_2$ *continued*

Conclusions

1. Inferring Conclusions and Resolving Discrepancies Compare your results for test tubes 1 and 2. Test tube 2 had only half as much hydrogen; was the pop test only half as loud? Explain your answer, referring to the balanced chemical equation for the pop test.

____

____

____

____

2. Inferring Conclusions and Resolving Discrepancies Consider your results for test tube 3. Did the hydrogen ignite? Can you explain why you observed what you did?

____

____

____

3. Inferring Conclusions and Resolving Discrepancies Is hydrogen gas more dense or less dense than air? (Hint: compare your results for test tubes 4 and 5. What was different about how they were treated, and how did this affect their pop test results?)

____

____

____

____

4. Organizing Conclusions Using your observations of the hydrogen gas in each trial, the generation reaction, and the pop tests, list as many chemical properties of hydrogen as you can. Explain what evidence you have for each one. (At least four properties were observable in the experiment.)

____

____

____

____

____

____

____

5. Organizing Conclusions Using your observations from this experiment, list as many physical properties of hydrogen as your can. Explain what evidence you have for each one. (At least four properties were observable in this experiment.)

____

____

____

____

____

____

Generating and Collecting H$_2$ *continued*

---

6. **Predicting Outcomes and Designing Experiments** Air is about 20% oxygen. Use this information to predict what ratio of air to hydrogen will provide the loudest pop test. Design an experiment that will test a range of proportions, including the one that you predicted would be the loudest. If your teacher approves your suggestions, test your predictions.

____

____

____

____

____

Testing Reaction Combinations of H_2 and O_2

Hydrogen, H_2, and oxygen, O_2, are two gases that react with each other in a very quick, exothermic (heat-producing) manner. The explosiveness of this reaction is greatest when hydrogen and oxygen are present in just the proper proportion. The reaction is used to power three of the rocket engines that carry the Space Shuttle into orbit.

In this experiment, you will generate hydrogen and oxygen and collect them using the water displacement method. You will test their explosive nature, first separately and then in reactions of different proportions of the two gases. After finding the most powerful reaction combination, you will use this reaction to launch a rocket across the room.

OBJECTIVES

- **Observe** the production of H_2 and O_2.

- **Use** the technique of water displacement to collect H_2 and O_2 gas.

- **Relate** chemical concepts such as stoichiometry, limiting reactant, activation energy, and thermodynamics to observations of chemical reactions.

- **Infer** a conclusion from experimental data.

- **Evaluate** experimental methods.

MATERIALS

- 1.0 M HCl solution
- 1.0 M KI solution
- 3% H_2O_2 solution
- 100 mL (or larger) beakers, 2
- collection bulb, calibrated
- micro H_2 generator
- plastic wash bottle

- micro O_2 generator
- petri dish, half
- piezoelectric sparker
- mossy zinc
- tongs or forceps
- tap water

Always wear safety goggles and a lab apron to protect your eyes and clothing. If you get a chemical in your eyes, immediately flush the chemical out at the eyewash station while calling to your teacher. Know the locations of the emergency lab shower and eyewash station and the procedures for using them.

 **Do not touch any chemicals.** If you get a chemical on your skin or clothing, wash the chemical off at the sink while calling to your teacher. Make sure you carefully read the labels and follow the precautions on all containers of chemicals that you use. If there are no precautions stated on the label, ask your teacher what precautions you should follow. Do not taste any chemicals or items used in the laboratory. Never return leftovers to their original containers; take only small amounts to avoid wasting supplies.

Call your teacher in the event of a spill. Spills should be cleaned up promptly, according to your teacher's directions.

PROCEDURE

1. Label the two beakers *water* and *spent solution*. Be sure both beakers are nearby as you proceed. Fill the beaker labeled *water* full of water and the petri dish three-fourths full of water.

2. The micro H$_2$ generator is a plastic vial containing several pieces of zinc, Zn, metal and capped with a lid and nozzle.
   Before continuing with the experiment, make sure both generator caps have a nozzle and that the nozzle is not plugged in any way.
   Remove the cap of the H$_2$ generator, and carefully add enough 1.0 M hydrochloric acid, HCl, to fill the vial to within 1 cm of the top. Replace the cap and set the H$_2$ generator in the petri dish. Observe the reaction, and record your observations.

Observations:

____

____

____

3. The micro O$_2$ generator is a plastic vial capped with a lid having a nozzle, as shown in **Figure A**. Remove the cap, and carefully add enough 3% hydrogen peroxide, H$_2$O$_2$, to fill the vial to within 1 cm of the top. Then add 20 drops of potassium iodide, KI, solution to serve as a catalyst. Replace the cap and set the O$_2$ generator in the petri dish beside the H$_2$ generator. Record your observations about the reaction in your lab notebook.

Figure A

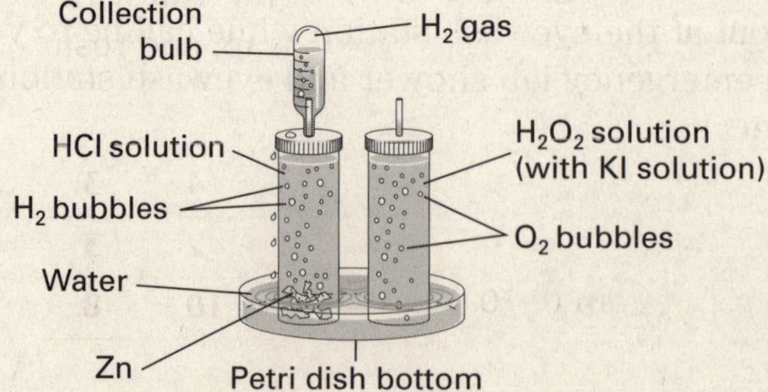

4. Fill the collection bulb completely with water from the petri dish or a plastic wash bottle. If you are using the petri dish, the best way is to squeeze the bulb tightly, invert it into the petri dish, and draw up as much water as possible. Then hold the bulb mouth-upward and squeeze out the remaining air. Without letting go of the bulb, invert it again into the petri dish and draw up the remainder of the water needed to fill the bulb.

5. To collect hydrogen gas, place the water-filled collection bulb from step **4** mouth-downward over the nozzle of the H_2 generator, as shown in **Figure A.** As soon as the bulb is completely filled with hydrogen, remove it from the nozzle and place a finger over the mouth of the bulb to prevent the collected gas from escaping.

6. Test the bulb of hydrogen gas by moving your finger aside and inserting the wire tip of the piezoelectric sparker into the bulb. **Keep the piezoelectric sparker away from the gas-generating vials. Hold the bulb securely. (Your teacher will tell you whether you should hold the bulb with your hand, with tongs, or with forceps.)** Pull the trigger of the piezoelectric sparker and observe. Record the loudness of the reaction (on a scale of 1 to 10) in the **Data Table.**

7. To collect oxygen gas, refill the bulb with water, place the bulb mouth-downward over the nozzle of the O_2 generator (you may need to gently agitate the vial). Repeat steps **5** and **6,** pop-testing the bulb with the piezoelectric sparker. Record the loudness of the reaction (on a scale of 1 to 10) in the **Data Table.**

8. Refill the bulb with water, and begin collecting another bulb of O_2, but when the bulb is about one-sixth full of O_2 (with five-sixths water), move the bulb from the O_2 generator to the H_2 generator, and continue collecting. When the bulb is filled with gas, you will have a combination that is 1 part oxygen and 5 parts hydrogen. Remove the bulb from the nozzle, test it as in step **6,** and record the relative loudness of the reaction in the **Data Table.**

9. Repeat step **8,** making the switch from the O_2 to the H_2 generator at different times so that you can test various proportions (2:4, 3:3, 4:2, and 5:1) to complete the **Data Table.** Determine the optimum (most explosive) combination.

10. If either generator reaction slows down so that it takes more than 1 min to fill the bulb with gas, lift off the bulb, uncap the vial, and decant the remaining liquid into the *spent solution* beaker. Refill the vial with fresh solution(s) from the appropriate bottle(s), replace the cap, and resume collecting gas.

DATA TABLE

Parts H_2	6	5	4	3	2	1	0
Parts O_2	0	1	2	3	4	5	6
Relative loudness (rated from 0–10)	**0**	**7**	**10**	**8**	**5**	**2**	**0**

11. Rocket launch (check with your teacher before performing this step):
Collect the optimum (loudest) mixture one more time, and instead of holding
onto the bulb, aim it outward at a target chosen by your teacher. **Make sure
nobody is in the line of fire.** Pull the trigger and launch your microrocket.
Can you think of ways to make your rocket go farther? Try them!

DISPOSAL

12. Clean all apparatus and your lab station. Return equipment to its proper
place. Dispose of chemicals and solutions in the containers designated
by your teacher. Do not pour any chemicals down the drain or in the
trash unless your teacher directs you to do so. Wash your hands thoroughly
before you leave the lab and after all work is finished.

Analysis

1. Relating Ideas Write a balanced equation for the reaction

  a. taking place inside the O$_2$ generator.

  b. taking place inside the H$_2$ generator.

____

____

2. Interpreting Ideas Which do you think will be used up first: the KI solution
in the O$_2$ generator or the Zn in the H$_2$ generator? Why?

____

____

____

3. Analyzing Methods What laboratory techniques did you use to help ensure
that the collection bulb contained only the desired gas?

____

____

____

____

____

Testing Reaction Combinations of H$_2$ and O$_2$ *continued*

4. Analyzing Results Make a bar graph of the relative loudness produced by your pop tests of the mixtures, with the loudness from 0 through 10 on the vertical axis and trials 1 through 7 on the horizontal axis.

Conclusions

1. Inferring Conclusions and Relating Ideas From your observations, state the relative combustibility of the following:

a. pure O$_2$

b. pure H$_2$

____

____

____

____

____

____

____

2. Resolving Discrepancies and Inferring Relationships Were there any reaction combinations that produced no reaction at all? Explain what happened.

____

____

____

____

3. Inferring Conclusions and Inferring Relationships What proportion of oxygen to hydrogen produced the most explosive reaction? Explain why that combination was the most explosive by referring to the balanced chemical equation for the reaction and the concepts of limiting reactants and maximum yields.

____

____

____

____

____

____

____

____

4. Predicting Outcomes Would your results be different if you performed the same experiment with the gases chilled? If your teacher approves, test your prediction. Explain your results.

____

____

____

____

____

____

____

____

Testing Reaction Combinations of H$_2$ and O$_2$ *continued*

5. Resolving Discrepancies and Relating Ideas Why don't O$_2$ and H$_2$ react as soon as they mix in the collection bulb? The individual O$_2$ and H$_2$ molecules are certainly colliding with one another. What role does the spark or flame play?

____

____

____

____

6. Evaluating Information Share your data with the rest of the class. Calculate the average relative loudness values for each mixture for the entire class, and make a bar graph of the results.

____

____

____

____

____

7. Analyzing Methods and Inferring Conclusions What methods did you attempt for making your rocket fly farther? Which methods worked best? Explain why they worked.

____

____

____

____

____

____

____

Testing Reaction Combinations of H_2 and O_2 *continued*

8. Applying Models The Space Shuttle carries 680 000 kg of fuel for its main engines. These engines use liquid hydrogen and liquid oxygen. Based on your results in this experiment, what is the most effective way to divide this mass? How much of it should be liquid oxygen, and how much liquid hydrogen?

Relative Solubility of Transition Elements

The transition elements are found in periods 4, 5, and 6 between groups 2 and 13 of the periodic table. As the atomic number increases across a row in this section of the table, no new valence electrons are added to the highest energy level. Instead, additional electrons are added to inner energy levels. As a result, the properties of the transition elements are similar across a row as well as down a column. In many instances, the properties of the elements in a given row are more alike than the properties of elements in a given column. This is not the case for nontransition elements.

In this experiment, you will evaluate the solubility of compounds of iron, copper, zinc, and silver in water and in acid. Iron, copper, and zinc are in the same row of the periodic table; copper and silver are in the same column. You will mix soluble ionic compounds of these transition elements with five different reactants and observe whether an insoluble compound (precipitate) forms. For those mixtures that form precipitates, you will investigate how the precipitate behaves in acid. After tabulating your findings, you will decide whether each new compound is soluble or insoluble in water and to what extent it dissolves in acid. You can then look for trends related to the positions of the elements in the periodic table.

OBJECTIVES

- **Observe** the reactions of some transition-element ions with five anions.

- **Deduce** the solubility of transition-element compounds in water and in acid.

- **Compare** the solubilities of elements in the same column and in the same row of the periodic table.

MATERIALS

- 24-well microplate or plastic sheet

- reactants
  0.1 M $K_4Fe(CN)_6$
  0.1 M KSCN
  0.1 M NaCl
  0.1 M Na_2SO_4
  0.2 M $(NH_4)_2C_2O_4$
  1.0 M HNO_3

- transition elements
  0.1 M $Cu(NO_3)_2$
  0.1 M $Fe(NO_3)_3$
  0.1 M $AgNO_3$
  0.1 M $Zn(NO_3)_2$

- thin-stemmed pipets or dropper bottles for solutions, 10

 **For this experiment, wear safety goggles, gloves, and a lab apron, to protect your eyes, hands, and clothing.** If you get a chemical in your eyes, immediately flush the chemical out at the eyewash station while calling to your teacher. Know the locations of the emergency lab shower and eyewash station and the procedures for using them.

Do not touch any chemicals. If you get a chemical on your skin or clothing, wash the chemical off at the sink while calling to your teacher. Make sure you carefully read the labels and follow the precautions on all containers of chemicals that you use. If there are no precautions stated on the label, ask your teacher what precautions you should follow. Do not taste any chemicals or items used in the laboratory. Never return leftovers to their original containers; take only small amounts to avoid wasting supplies.

Call your teacher in the event of a spill. Spills should be cleaned up promptly, according to your teacher's directions.

PROCEDURE

1. Construct a full-page grid similar to the **Solubility Chart** in Analysis Question **1**. Place a microplate or plastic sheet over the grid.

2. Place two drops of each transition-element solution in the appropriate well on the microplate. These solutions contain positive ions. Be careful; silver nitrate, $AgNO_3$, will stain skin and clothing. There should be five boxes in a row for each transition element. Add two drops of each of the reactants, which contain negative ions, to each of the transition-element solutions. **Do not allow the dropper to touch the transition-element solutions.**

3. Immediately after adding each reactant solution to each transition solution, record your observations in the section of the **Data Table** labeled *0 min*. If no reaction occurs, write *NR*. If a precipitate forms, record its color.

4. Wait 5 min, and then record any changes in the mixtures in the section of the **Data Table** labeled *5 min*. If no change occurs, write *same*.

5. To each mixture in which a precipitate formed, add one to four drops of 1.0 M nitric acid. Write *clear* in the *acid* section of the **Data Table** if the precipitate disappeared. Write *PC* if the precipitate partially disappeared and *same* if there was no change. Record any color change.

   CAUTION Nitric acid, HNO_3, is corrosive and caustic. Avoid contact with eyes and skin. If any should spill on you, immediately flush the area with water and notify your teacher.

DISPOSAL

6. Clean all apparatus and your lab station. Return equipment to its proper place. Dispose of chemicals and solutions in the containers designated by your teacher. Do not pour any chemicals down the drain or into the trash unless your teacher directs you to do so. Wash your hands thoroughly before you leave the lab and after all work is finished.

Relative Solubility of Transition Elements *continued*

DATA TABLE

Transition ions		Reactant ions				
		Cl^-	$C_2O_4^{2-}$	SO_4^{2-}	SCN^-	$Fe(CN)_6^{4-}$
Cu^{2+}	0 min					
	5 min					
	acid					
Fe^{3+}	0 min					
	5 min					
	acid					
Ag^+	0 min					
	5 min					
	acid					
Zn^{2+}	0 min					
	5 min					
	acid					

Relative Solubility of Transition Elements *continued*

Analysis

1. **Organizing Data and Analyzing Results** Study your **Data Table** carefully. Then complete the **Solubility Chart** below. For each combination of positive transition-element ions and negative reactant ions, write *S* if the mixture is soluble in water (no reaction) and *PS* if it is partially soluble in water. Write *SA* if the mixture is soluble in acid but not in water and *PSA* if it is partially soluble in acid but not in water. Write *I* if the mixture is insoluble in water and acid. Assume a mixture is partially soluble in water if it was clear at 0 min but became cloudy after 5 min. Assume a mixture is partially soluble in acid if the precipitate partially dissolved after the addition of HNO_3.

SOLUBILITY CHART

Transition ions	Reactants				
	Cl^-	$C_2O_4^{2-}$	SO_4^{2-}	SCN^-	$Fe(CN)_6^{4-}$
Cu^{2+}					
Fe^{3+}					
Ag^+					
Zn^{2+}					

2. **Analyzing Results** Which transition-element ion(s) formed the greatest number of soluble compounds with the five negative ions used in this experiment?

____

____

____

____

3. **Analyzing Results** Which transition-element ion(s) formed the smallest number of soluble compounds with the five negative ions used in this experiment?

____

____

____

____

Relative Solubility of Transition Elements *continued*

4. Analyzing Results Copper and silver are in the same vertical column of the periodic table. Iron, copper, and zinc are in the same horizontal row. According to your **Solubility Chart,** are the solubilities more similar across the row, down the column, or neither?

____

____

____

5. Analyzing Information Use your **Data Table** and **Solubility Chart** to identify each of the following unknown ions.

a. Forms a white precipitate with Cl^-

____

b. Forms a precipitate with at least four reactant ions

____

c. Two reactant ions that can distinguish Ag^+ from the other three transition-element ions

____

d. The reactant ion that forms the most insoluble transition-metal compounds

____

Conclusion

1. Predicting Outcomes Notice the position of cadmium in the periodic table. On the basis of your data on zinc, copper, and silver, predict the overall solubility of cadmium.

____

____

____

____

____

Reacting Ionic Species in Aqueous Solution

When ionic solids dissolve in water, they dissociate into positive *cations* and negative *anions*. If two solutions containing dissolved ionic solids are mixed together, new combinations of cations and anions are possible. Sometimes a new combination of ions is not soluble in water, and a precipitate (solid) forms. For example, silver nitrate in solution dissociates into Ag^+ and NO_3^- ions, and KCl in solution breaks up into K^+ and Cl^- ions. If the two solutions are mixed, the positive Ag^+ ions combine with the negative Cl^- ions to form an insoluble precipitate of AgCl. The *complete ionic* equation for the reaction is the following.

$$Ag^+(aq) + NO_3^-(aq) + K^+(aq) + Cl^-(aq) \rightarrow AgCl(s) + K^+(aq) + NO_3^-(aq)$$

Because the K^+ ion and NO_3^- ion do not take part in the reaction, they are called spectator ions. Subtraction of the spectator ions from both sides of the equation produces the *net ionic* equation for the reaction.

$$Ag^+(aq) + Cl^-(aq) \rightarrow AgCl(s)$$

In this experiment, you will mix six different ionic solutions, two at a time, and observe which combinations form precipitates. You will also determine which pairs of ions form precipitates, and you will write net ionic equations for the reactions.

OBJECTIVES

Organize experimental variables to study them one at a time.

Recognize a precipitation reaction.

Formulate net ionic equations for precipitation reactions from experimental data.

MATERIALS

- $BaCl_2$, 0.1 M
- $Ba(NO_3)_2$, 0.1 M
- gloves
- lab apron
- $Na_2C_2O_4$, 0.1 M
- NaCl, 0.1 M
- $NaNO_3$, 0.1 M
- Na_2SO_4, 0.1 M
- plastic sheet
- safety goggles

Always wear safety goggles and a lab apron to protect your eyes and clothing. If you get a chemical in your eyes, immediately flush the chemical out at the eyewash station while calling to your teacher. Know the location of the emergency lab shower and eyewash station and the procedures for using them.

▌Reacting Ionic Species in Aqueous Solution *continued*

 **Do not touch any chemicals.** If you get a chemical on your skin or clothing, wash the chemical off at the sink while calling to your teacher. Make sure you carefully read the labels and follow the precautions on all containers of chemicals that you use. If there are no precautions stated on the label, ask your teacher what precautions to follow. Do not taste any chemicals or items used in the laboratory. Never return leftovers to their original container; take only small amounts to avoid wasting supplies.

Call your teacher in the event of a spill. Spills should be cleaned up promptly, according to your teacher's directions.

Acids and bases are corrosive. If an acid or base spills onto your skin or clothing, wash the area immediately with running water. Call your teacher in the event of an acid spill. Acid or base spills should be cleaned up promptly.

Procedure

1. Put on safety goggles, gloves, and a lab apron.

2. Decide on the order in which you will perform the tests before you begin. Record your observations in **Table 1** after you have performed each test. For each possible combination, indicate *PPT* if a precipitate forms and *clear* if no precipitate forms.

3. Place a drop or two of one of the solutions on the plastic sheet. Add a drop or two of the second solution. Do not allow the tip of the dropper to touch the drops on the plastic sheet. Record your result.

4. Choose a clean spot on the plastic and repeat step 2 for another pair of solutions. Repeat for all possible combinations indicated in **Table 1,** and record your results.

5. Clean all apparatus and your lab station. Return equipment to its proper place. Dispose of chemicals and solutions in the containers designated by your teacher. Do not pour any chemicals down the drain or in the trash unless your teacher directs you to do so. Wash your hands thoroughly before you leave the lab and after all work is finished.

TABLE 1: TEST RESULTS FOR EACH POSSIBLE COMBINATION

	Na^+, Cl^-	Na^+, NO_3^-	Na^+, $C_2O_4^{2-}$	Na^+, SO_4^{2-}	Ba^{2+}, Cl^-
Ba^{2+}, NO_3^-	clear	clear	PPT	PPT	clear
Ba^{2+}, Cl^-	clear	clear	PPT	PPT	
Na^+, SO_4^{2-}	clear	clear	clear		
Na^+, $C_2O_4^{2-}$	clear	clear			
Na^+, NO_3^-	clear				

Name _______________________________________ Class ________________ Date ______________

Analysis

1. **Analyzing Information** When a precipitate forms, you can assume that ions have combined to form an insoluble compound. For example, if $CaCl_2$ is mixed with Na_2SO_4, a precipitate forms. The possible combinations that could have produced the precipitate are $CaSO_4$ and $NaCl$. Write the possible formulas for each precipitate that you observed.

2. **Analyzing Information** Compare the possible precipitates listed in **item 1** with the chemicals you used. Identify the formula from each pair of possible precipitates that you think is most likely to be the precipitate. Explain your answer.

   ____

   ____

   ____

3. **Organizing Conclusions** Write a balanced *complete ionic* equation for each precipitate formed. Use the *complete* ionic equation in the Introduction as a guide.

4. **Organizing Conclusions** Write a *net ionic* equation for each reaction in which a precipitate is formed. Use the net ionic equation in the Introduction as a guide.

   ____

   ____

Conclusions

Use the information in the following table to answer **Items 1 through 5.** The table gives the solubilities of the listed compounds in moles of anhydrous compound that can be dissolved in 1 L of water at 20°C. According to the table, $NaNO_3$, $NaCl$, and $AgNO_3$ are soluble ionic compounds. Their aqueous solutions contain hydrated ions that can be present in high concentrations at room temperature.

TABLE 2: SOLUBILITY DATA

Compound	Formula	Solubility at 20°C (mol/L)
silver nitrate	$AgNO_3$	13
silver chloride	$AgCl$	1×10^{-5}
sodium nitrate	$NaNO_3$	10.3
sodium chloride	$NaCl$	6.2

1. **Organizing Information** Suppose that at room temperature you mix fairly concentrated solutions of NaCl and $AgNO_3$. What four ionic species are present in this mixture of two ionic compounds?

____

2. **Analyzing Information** When the fairly concentrated solutions of NaCl and $AgNO_3$ are mixed, what compound is likely to precipitate out of solution? Explain why.

____

____

____

____

3. **Organizing Conclusions** Write the overall ionic equation for the precipitation reaction described in **Conclusion Item 2.**

____

4. **Organizing Conclusions** The overall ionic equation in **Item 3** shows that the $Na^+(aq)$ and $NO_3^-(aq)$ ions take no part in the reaction. What is the name given to ions that take no part in a chemical reaction?

____

5. **Organizing Conclusions** Write the net ionic equation for the reaction between NaCl and $AgNO_3$.

Colored Precipitates

Binary ionic compounds exhibit a variety of properties. Some dissolve easily in water, while others are insoluble. Some are brightly colored compounds; many others are white. You can use characteristic properties of known substances to determine the identity of an unknown by comparing the properties of a known substance with those of the unknown.

In this experiment, you will carry out some double-displacement reactions and observe the characteristic colors of precipitates. You will use your experimental data to identify unknown substances.

OBJECTIVES

- **Observe** displacement reactions in which precipitates are formed.

- **Compare** chemical and physical properties of substances.

- **Relate** observations to the identification of unknown solutions.

- **Infer** a conclusion from experimental data.

MATERIALS

- 0.1 M $CoCl_2$
- 0.1 M $CuCl_2$
- 0.1 M $FeCl_3$
- 0.1 M NaOH
- 0.1 M $NiCl_2$

- dropper bottles, 8
- 8-well flat-bottom strip or 8 small test tubes
- mystery solutions *1*, *2*, and *3*
- thin-stemmed pipets, 8

Always wear safety goggles and a lab apron to protect your eyes and clothing. If you get a chemical in your eyes, immediately flush the chemical out at the eyewash station while calling to your teacher. Know the locations of the emergency lab shower and eyewash station and the procedures for using them.

Do not touch any chemicals. If you get a chemical on your skin or clothing, wash the chemical off at the sink while calling to your teacher. Make sure you carefully read the labels and follow the precautions on all containers of chemicals that you use. If there are no precautions stated on the label, ask your teacher what precautions you should follow. Do not taste any chemicals or items used in the laboratory. Never return leftovers to their original containers; take only small amounts to avoid wasting supplies.

Call your teacher in the event of a spill. Spills should be cleaned up promptly, according to your teacher's directions.

Colored Precipitates *continued*

PROCEDURE

1. Label a sheet of paper to identify the solutions, as shown in **Figure A.** Place it under the 8-well strip.

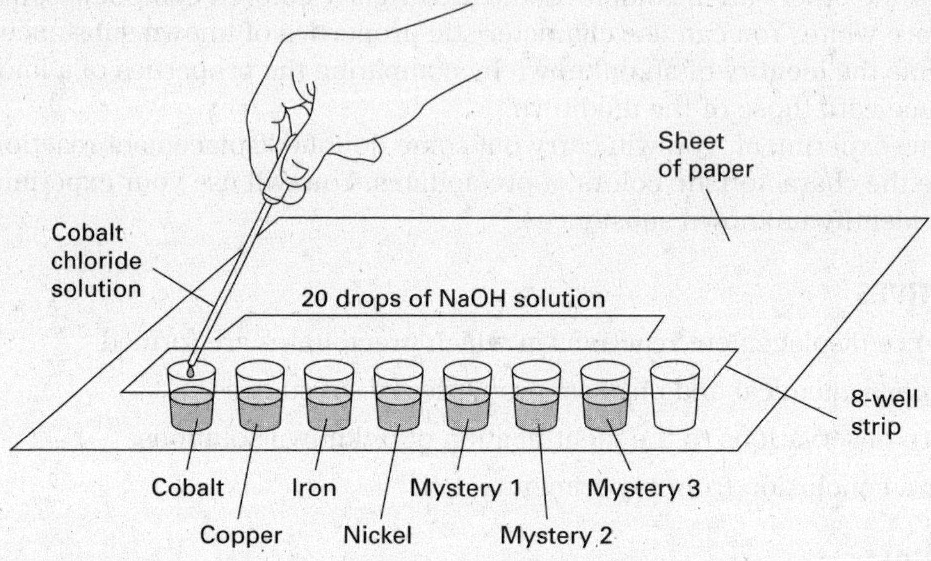

Figure A

2. With a thin-stemmed pipet, place 20 drops of sodium hydroxide, NaOH, in the first seven of the eight wells in the strip. (If you are using test tubes, use 40 drops.) For best results, all the drops should be the same size.

3. To the first well, add 5 drops of cobalt(II) chloride solution, $CoCl_2$. (If using test tubes, add 10 drops of each metal chloride to each tube.) To the second well, add 5 drops of copper(II) chloride solution, $CuCl_2$. To the third, add 5 drops of iron(III) chloride, $FeCl_3$. To the fourth, add 5 drops of nickel(II) chloride, $NiCl_2$. Observe what happens. Record your observations and the physical properties of the substances formed in the **Data Table.**

4. To the fifth well, add 5 drops of mystery solution *1*. (If using test tubes, add 10 drops of each of the mystery solutions.) To the sixth well, add 5 drops of mystery solution *2*. To the seventh well, add 5 drops of mystery solution *3*. Observe what happens. Record your observations about the kinds of reactions you see and the physical properties of the substances formed.

DISPOSAL

5. Clean all apparatus and your lab station. Return equipment to its proper place. Dispose of chemicals and solutions in the containers designated by your teacher. Do not pour any chemicals down the drain or in the trash unless your teacher directs you to do so. Wash your hands thoroughly before you leave the lab and after all work is finished.

Colored Precipitates *continued*

DATA TABLE

Well number	1	2	3	4	5	6	7
Label	$CoCl_2$	$CuCl_2$	$FeCl_3$	$NiCl_2$	Mystery solution 1	Mystery solution 2	Mystery solution 3
Observations in step 3							
Observations in step 4							

Analysis

1. **Organizing Ideas** The reactions in this lab are all double-displacement reactions. Write chemical equations for the reactions occurring in the first four mixtures.

____

____

____

____

____

____

2. **Organizing Ideas** A net ionic equation is one that includes only those dissolved ions that undergo a reaction. For example, to write a net ionic equation for the reaction of HCl with NaOH, take the following steps:

- Write the balanced equation showing the complete formulas for the reactants and products.

$$HCl(aq) + NaOH(aq) \rightarrow NaCl(aq) + H_2O(l)$$

- Rewrite the equation with all ions separated, or dissociated. Formulas for solid precipitates, gases, or liquids should be written as undissociated compounds.

$$H^+(aq) + Cl^-(aq) + Na^+(aq) + OH^-(aq) \rightarrow Na^+(aq) + Cl^-(aq) + H_2O(l)$$

Colored Precipitates *continued*

- Cross out any ions that appear in the same form on both sides of the equation (in this case, Na^+ and Cl^-).

$$H^+(aq) + \cancel{Cl^-(aq)} + \cancel{Na^+(aq)} + OH^-(aq) \rightarrow \cancel{Na^+(aq)} + \cancel{Cl^-(aq)} + H_2O(l)$$

- Rewrite the equation, omitting any ions that were crossed out.

$$H^+(aq) + OH^-(aq) \rightarrow H_2O(l)$$

This is the net ionic equation that includes only those ions undergoing a chemical change. Write a net ionic equation for each of the reactions that occurred in the first four mixtures.

____

____

____

____

3. Analyzing Results Would the reactions have been different if potassium hydroxide, KOH, had been used instead of sodium hydroxide? Explain your answer. Hint: Write the net ionic equation using KOH(aq) instead of NaOH(aq).

____

____

____

____

4. Inferring Conclusions Which of the net ionic equations that you wrote in item **2** describe the reactions of the three mystery solutions? Write the equations, and explain the reasons you chose them.

____

____

____

____

Colored Precipitates *continued*

5. **Relating Ideas** All the reactants in these reactions are solutions of ionic compounds. Do you think the characteristic colors of the reactants and the products are caused by the positive or the negative ions in the compounds? Explain your reasoning.

____

____

____

____

____

____

6. **Relating Ideas** How many drops of NaOH are needed to react completely with five drops of each of the metal chloride solutions if all the drops are the same size? (Hint: Use the balanced chemical equations, and recall that the concentrations of all the reactants are the same.)

____

____

____

____

____

____

7. **Predicting Outcomes** Use your answer to item **6** to decide which reactant, NaOH or metal chloride, would be left over in each well after the reaction. How much of the excess reactant would be present?

____

____

____

____

____

Colored Precipitates *continued*

Conclusions

1. **Analyzing Results** Can you tell if the mystery solutions had concentrations of positive ions that were greater than, less than, or the same as the concentrations of the known solutions you tested? Explain your reasoning.

____

____

____

____

____

2. **Designing Experiments** Can you think of a way to use the products of these reactions to make an estimate of the concentration of the mystery solutions? Write a procedure for making this determination. If your teacher approves, test your procedure.

____

____

____

____

____

____

____

____

Colored Precipitates *continued*

3. **Designing Experiments** The 0.1 M NaOH solution used as a reactant in this reaction has a pH of 13.0. The mathematical relationship of pH to OH⁻ concentration in moles per liter is given by the following equation:

$$pH = 14.00 + \log [OH^-]$$

Describe how you could use a pH meter to estimate the concentration of the positive ions in the mystery solutions.

____

____

____

____

____

____

____

____

____

____

____

Percentage of Acetic Acid in Vinegar

When sweet apple cider is fermented in the absence of oxygen, the product is an acid, vinegar. Most commercial vinegars are made by fermentation, but some, such as the white vinegar you will use in this experiment, are obtained by the dilution of 100% acetic acid. The usual mass percentage of acetic acid in vinegar is between 4.0% and 5.5% regardless of how it is produced.

The quantity of acetic acid in a sample of vinegar may be found by titrating the sample against a standard basic solution. In this experiment, you will titrate with pipets to determine what volume of sodium hydroxide solution of known molarity is needed to neutralize a measured quantity of a vinegar. From your results and the molarity of sodium hydroxide, NaOH, you can calculate the molarity of the vinegar.

OBJECTIVES

- **Determine** the end point of an acid-base titration.

- **Observe** accurately the quantities of solution in pipets.

- **Calculate** the molarity of vinegar, using experimental data.

- **Calculate** the percentage of acetic acid in vinegar.

MATERIALS

- 10 mL graduated cylinder
- 24-well plate
- thin-stemmed pipets, 2
- phenolphthalein indicator, 1 dropper bottle
- HCl, for standardizing NaOH
- NaOH solution, standardized
- white vinegar

Always wear safety goggles and a lab apron to protect your eyes and clothing. If you get a chemical in your eyes, immediately flush the chemical out at the eyewash station while calling to your teacher. Know the locations of the emergency lab shower and eyewash station and the procedures for using them.

Do not touch any chemicals. If you get a chemical on your skin or clothing, wash the chemical off at the sink while calling to your teacher. Make sure you carefully read the labels and follow the precautions on all containers of chemicals that you use. If there are no precautions stated on the label, ask your teacher what precautions you should follow. Do not taste any chemicals or items used in the laboratory. Never return leftovers to their original containers; take only small amounts to avoid wasting supplies.

Call your teacher in the event of a spill. Spills should be cleaned up promptly, according to your teacher's directions.

Percentage of Acetic Acid in Vinegar *continued*

Never put broken glass in a regular waste container. Broken glass should be disposed properly.

PROCEDURE

1. To calibrate the pipet you will use for the vinegar, put 5.0 mL of water into the 10 mL graduated cylinder. Read the exact volume of water in the cylinder, and record this volume as the initial reading for the acid pipet, trial l, in the **Calibration Data Table.** Fill the thin-stemmed pipet with water from the faucet. Hold the pipet vertically, and transfer 20 drops of water from the pipet to the graduated cylinder. Read and record the new volume as the final volume for trial 1. Without emptying the cylinder, add 20 more drops and record the volume for this second trial. Repeat for a third trial. Label the pipet *Acid*.

2. Repeat step **1** for the second pipet, which will be the base pipet. Record the data, and label this pipet *NaOH*.

CALIBRATION DATA TABLE

| | Readings in Graduated Cylinder (mL) | | | |
| | Acid pipet | | Base pipet | |
Trial	Initial	Final	Initial	Final
1				
2				
3				

Total volume of drops from acid pipet: _______________________

Average volume of each drop of acid: _______________________

Total volume of drops from base pipet: _______________________

Average volume of each drop of base: _______________________

3. Hold the pipets vertically each time you use them. Drain the water from the *Acid* pipet, and fill it with vinegar. Add 20 drops of vinegar and 1 drop of phenolphthalein to each of 3 wells of the 24-well plate.

4. Drain the water from the *NaOH* pipet, and fill it with NaOH. To one of the wells, add the NaOH solution drop by drop from a vertical pipet, swirling the plate gently after each drop. Continue until the pink color persists for 30 s. Record the number of drops added in the **Titration Data Table.**

5. Repeat the titration in step **4** with the two other 20-drop samples of vinegar. Perform each titration in the same way, titrating to the same shade of pink each time. Refill the *NaOH* hydroxide pipet, if necessary. Record your results.

Percentage of Acetic Acid in Vinegar *continued*

DISPOSAL

6. Clean all apparatus and your lab station. Return equipment to its proper place. Dispose of chemicals and solutions in the containers designated by your teacher. Do not pour any chemicals down the drain or in the trash unless your teacher directs you to do so. Wash your hands thoroughly before you leave the lab and after all work is finished.

TITRATION DATA TABLE

Number of Drops Added to Well		
Trial	Vinegar	NaOH
1		
2		
3		

Analysis

1. Organizing Data Calculate the volumes of vinegar and NaOH used for each of the three trials.

Volume of vinegar = ____

Volume of NaOH:

Trial 1: ____

Trial 2: ____

Trial 3: ____

2. Organizing Data From the molarity of the standardized NaOH solution you used, determine the number of moles of NaOH used in each of the three trials.

Molarity NaOH = ____

Trial 1: ____

Trial 2: ____

Trial 3: ____

3. Organizing Ideas Write the balanced equation for the reaction between vinegar and sodium hydroxide. (Hint: The formula for acetic acid is CH_3COOH.)

____

Percentage of Acetic Acid in Vinegar *continued*

4. Organizing Data Use the results of your calculations in item **2** and the mole ratio in the equation in item **3** to determine the number of moles of base used to neutralize the vinegar (acid) in each trial.

Trial 1: ____

Trial 2: ____

Trial 3: ____

5. Organizing Data Use the moles of acid calculated in item **4** and the volumes of the acid used for each trial to calculate the molarities of vinegar for the three trials.

Trial 1: ____

Trial 2: ____

Trial 3: ____

6. Organizing Data Calculate the average molarity of the vinegar.

7. Organizing Ideas Use the periodic table to calculate the mass of 1 mol of acetic acid, CH_3COOH.

8. Organizing Data Use the average molarity for your vinegar sample to determine the mass of CH_3COOH in 1 L of vinegar.

9. Organizing Conclusions Assume that the density of vinegar is very close to 1.00 g/mL so that the mass of 1 L of vinegar is 1000 g. Calculate the percentage of acetic acid in your vinegar sample. (Hint: The mass of acetic acid in 1 L of vinegar, calculated in item **8,** divided by the total mass of vinegar in a liter and multiplied by 100 gives the percentage of acetic acid in vinegar.)

Conclusions

1. Applying Conclusions Why is it important for a company manufacturing vinegar to regularly check the molarity of its product?

____

____

____

____

____

2. Analyzing Methods What was the purpose of the phenolphthalein? Could you have titrated the vinegar sample without the phenolphthalein?

____

____

____

____

3. Analyzing Methods Why was it important to hold the pipets vertically each time you used them?

____

____

____

4. Evaluating Data Share your data with other lab groups, and calculate a class average for the molarity of the vinegar. Compare the class average with your results, and calculate the average deviation and the standard deviation.

❚ Percentage of Acetic Acid in Vinegar *continued*

---

5. **Designing Experiments** What possible sources of error can you identify in this procedure? If you can think of ways to eliminate them, ask your teacher to approve your plan, and run the procedure again.

____

____

____

6. **Relating Ideas** Explain the difference between the equivalence point and the end point. How are they different? Can they be the same?

____

____

____

____

____

____

____

____

Determination of Vitamin C in Fruit Juices

Vitamin C is an important nutrient in the human diet. It is essential for preventing the disease called scurvy and for maintaining good health. Vitamin C is required daily because it is water soluble and cannot be stored by the body. Fortunately, vitamin C is abundant in foods, especially in citrus fruits. In this experiment, you will determine the concentration of vitamin C in three fruit juices by comparing them with a solution of vitamin C having a known concentration. To do so, you will take advantage of the ability of vitamin C to act as a reducing agent.

In cells, vitamin C is involved in a variety of oxidation-reduction reactions. When it reduces other molecules, such as iodine, I_2, it becomes oxidized. You will add iodine solution to a vitamin C solution until a starch indicator turns blue. The oxidation reaction is complex but can be summarized in this way:

$$\text{vitamin C} + I_2 \rightarrow 2I^- + \text{oxidized vitamin C}$$

When all of the vitamin C has been oxidized, the iodine is available to react with starch to form a blue color.

$$I_2 + \text{starch} \rightarrow \text{blue color}$$

In this experiment, you will determine the amount of I_2 needed to oxidize a sample of standard vitamin C solution and then compare it with the amount of iodine needed to oxidize a sample of citrus juice having the same volume. From your data you will calculate the concentration in mg/mL of vitamin C in the fruit juice.

OBJECTIVES

- **Determine** the amounts of iodine needed to oxidize the vitamin C in samples of fruit juice.

- **Compare** the amounts of iodine used in the fruit juice with the amount needed to oxidize a standard solution of vitamin C.

- **Calculate** the concentrations of vitamin C in the samples of fruit juice.

MATERIALS

- 24-well microplate
- thin-stemmed pipets, 4
- 1% starch solution
- dropper bottles, 3
- iodine solution
- fruit juices, 3 different types
- vitamin C solution, 1 mg/mL

Determination of Vitamin C in Fruit Juices *continued*

 **Always wear safety goggles and a lab apron to protect your eyes and clothing.** If you get a chemical in your eyes, immediately flush the chemical out at the eyewash station while calling to your teacher. Know the locations of the emergency lab shower and eyewash station and the procedures for using them.

Do not touch any chemicals. If you get a chemical on your skin or clothing, wash the chemical off at the sink while calling to your teacher. Make sure you carefully read the labels and follow the precautions on all containers of chemicals that you use. If there are no precautions stated on the label, ask your teacher what precautions you should follow. Do not taste any chemicals or items used in the laboratory. Never return leftovers to their original containers; take only small amounts to avoid wasting supplies.

Call your teacher in the event of a spill. Spills should be cleaned up promptly, according to your teacher's directions.

PROCEDURE

1. Put five drops of the standard vitamin C solution into each of three wells in the first row of the microplate, as shown in **Figure A.**

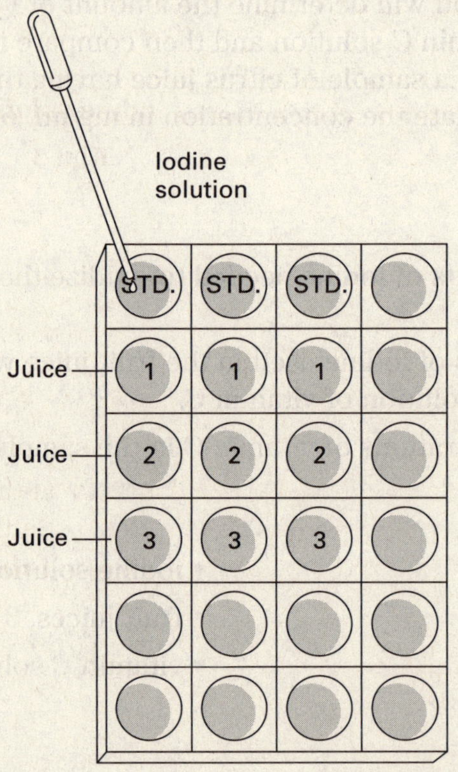

Figure A

2. Put five drops of fruit juice *1* into each of three wells in the second row of the plate.

Determination of Vitamin C in Fruit Juices *continued*

3. Put five drops of fruit juice *2* into each of three wells in the third row of the plate. Repeat this procedure for juice *3*, placing it in the wells in the fourth row.

4. Add one drop of starch solution to each well.

5. Count the drops as you add the iodine solution dropwise to the first of the vitamin C wells in the first row. Continue adding iodine until the solution in the well stays purple-blue for at least 15 s. Record the number of drops in the **Data Table.**

6. Repeat step **5** for the two other samples of standard vitamin C, and then do the same for each of the juice solutions in the wells. Record all your results in the **Data Table.**

DISPOSAL

7. Clean all apparatus and your lab station. Return equipment to its proper place. Dispose of chemicals and solutions in the containers designated by your teacher. Do not pour any chemicals down the drain or into the trash unless your teacher directs you to do so. Wash your hands thoroughly before you leave the lab and after all work is finished.

DATA TABLE

Solution	Drops of iodine solution			
	Trial 1	Trial 2	Trial 3	Average
Vitamin C				
Juice *1*				
Juice *2*				
Juice *3*				

Analysis

1. Organizing Data Determine the average number of drops of iodine solution required to oxidize the vitamin C in the standard solution and in the three juice samples. Record these averages in the **Data Table.**

2. **Organizing Data** In item **1,** you calculated the average number of drops of
iodine needed to oxidize the vitamin C in the standard vitamin C sample and
in the three juices. Recall that the concentration of the solution of vitamin C
is 1 mg/mL. You determined the number of iodine drops needed to oxidize five
drops of vitamin C solution. If you assume that all the drops delivered by the
four pipets were the same size, then you can calculate the concentrations of
vitamin C in the three fruit juices. The ratio of the concentration of vitamin C
in fruit juice *1* to the concentration of the standard vitamin C solution is equal
to the ratio of the drops of iodine solution used to titrate them.

Conclusions

1. **Inferring Conclusions** Which of the juices has the highest concentration of
vitamin C per milliliter of juice?

____

____

____

____

____

2. **Applying Conclusions** If the minimum daily requirement for vitamin C is
0.060 g/day, what is the minimum amount of juice you must drink each day?
(Assume that the juice is your only source of vitamin C.)

Clock Reactions

Many medications are advertised as having a timed-release action. When the medication is ingested, it becomes active in the body over time. Developers of pharmaceuticals can control how quickly a drug enters the bloodstream by controlling the rate at which the drug is digested in the stomach. In this experiment you will look at how temperature, surface area, and pH affect the rate at which an effervescent antacid tablet reacts with a solvent.

OBJECTIVES

- **Observe** chemical processes and interactions.

- **Compare** chemical reactions by measuring reaction rates.

- **Relate** reaction rate concepts to observations.

- **Infer** a conclusion from experimental data.

MATERIALS

- effervescent antacid tablets, 24
- 0.10 M HCl, 60 mL
- ice (optional)
- hot plate (optional)
- 50 mL beaker
- 25 mL graduated cylinder
- mortar and pestle

- scupula
- stirring rod
- timing device or stopwatch
- test tubes, 4
- test tube holder and rack
- thermometer, nonmercury
- waxed paper, 6 × 6 in.

Always wear safety goggles and a lab apron to protect your eyes and clothing. If you get a chemical in your eyes, immediately flush the chemical out at the eyewash station while calling to your teacher. Know the locations of the emergency lab shower and eyewash station and the procedures for using them.

Do not touch any chemicals. If you get a chemical on your skin or clothing, wash the chemical off at the sink while calling to your teacher. Make sure you carefully read the labels and follow the precautions on all containers of chemicals that you use. If there are no precautions stated on the label, ask your teacher what precautions you should follow. Do not taste any chemicals used in the laboratory. Never return leftovers to their original containers; take only small amounts to avoid wasting supplies.

Call your teacher in the event of a spill. Spills should be cleaned up promptly, according to your teacher's directions.

Clock Reactions *continued*

PROCEDURE

1. Fill four test tubes each with 15 mL of cold water from the tap. Record the temperature of the water in the **Data Table.**

2. Get three effervescent tablets. Break two of the tablets in half. Break one of the halves in half again.

3. Add a quarter of a tablet to the first test tube. Start timing, and record the reaction time in the **Data Table.**

4. Repeat step **3** three times, using a half tablet in one test tube, three-quarters of a tablet in another test tube, and a whole tablet in a fourth test tube. Record the reaction times for each test tube in the **Data Table.** Pour the solutions down the drain, and clean the test tubes for the next set of tests.

5. Repeat steps **1–4** with water at room temperature (about 25°C).

6. Repeat steps **1–4** with hot water from the tap.

7. Use room temperature water this time. Repeat steps **1–4,** but use a mortar and pestle to crush each of the samples. Place the powder on the waxed paper, and use a scupula to add the powder to the test tube quickly.

8. Repeat steps **1–4,** but crush each sample, use room temperature water in the test tubes, and stir the mixture.

9. Repeat steps **1–4,** but crush each sample, use cold water in the test tubes, and stir the mixture.

10. Repeat steps **1–4,** but crush each sample, use hot water in the test tubes, and stir the mixture.

11. Repeat steps **1–4,** but crush each sample, use 0.1 M HCl instead of water in the test tubes, and stir the mixture.

DISPOSAL

12. Clean all apparatus and your lab station. Return equipment to its proper place. Dispose of chemicals and solutions in the containers designated by your teacher. Do not pour any chemicals down the drain or in the trash unless your teacher directs you to do so. Wash your hands thoroughly before you leave the lab and after all work is finished.

Clock Reactions *continued*

DATA TABLE

	Solvent temperature (°C)	Time(s)			
		1/4 tablet	1/2 tablet	3/4 tablet	whole tablet
Cold water					
Room temperature water					
Hot water					
Crushed tablet, room temperature water					
Crushed tablet, room temperature water, stirred					
Crushed tablet, cold water, stirred					
Crushed tablet, hot water, stirred					
Crushed tablet, 0.1 M HCl, stirred					

Analysis

1. Analyzing Results Which mixture had the fastest reaction time? Which mixture had the slowest reaction time?

____

____

____

____

2. Analyzing Methods What was the effect of increased temperature on the reaction rate?

____

____

Clock Reactions *continued*

3. Evaluating Conclusions Which of the following reaction rate variables is being tested in this experiment: temperature, catalyst, concentration, surface area, or nature of reactants?

____

____

____

4. Inferring Conclusions Effervescent tablets generally contain a carbonate compound that decomposes to produce CO_2. Ask your teacher for the product label to determine the active ingredient in your sample. Write an equation for the reaction that produces CO_2.

____

____

____

Conclusions

1. Evaluating Methods What are some likely sources of error in this experiment? If you can think of ways to eliminate them, ask your teacher to approve your suggestion, and run more trials.

____

____

____

____

____

____

2. Predicting Outcomes and Designing Experiments Would all brands of effervescent tablets show the same reaction rates? Design an experiment to test your answer. If your teacher approves, perform the experiment and record your results.

____

____

____

Clock Reactions *continued*

3. Formulating Conclusions You are working for a pharmaceutical company. Your job is to write the labeling for a bottle of indigestion relief tablets. Suggest directions for the label to help customers obtain the fastest relief.

____

____

____

____

Equilibrium

Some chemical reactions run to completion and are commonly referred to as *end reactions*. These are reactions in which a product is essentially undissociated, given off as a gas, or precipitated. Many other reactions do not run to completion. The products of these reactions do not leave the field of action but react with each other to reform the original reactants. An equilibrium is established in which the forward and reverse actions continue at equal rates so that there is no net change in the quantities of reactants or products.

A reaction at equilibrium is affected by both concentration and temperature. Le Châtelier's principle states that a system in equilibrium tends to shift to relieve any stress placed on it by a change in concentration or temperature. For example, the addition of more reactants will shift an equilibrium in the direction of producing more products, thus relieving the stress on the reactants. In this experiment, you will investigate some systems at equilibrium and interpret your observations in terms of Le Châtelier's principle.

OBJECTIVES

Observe color changes in solutions as indications of shifts in equilibrium.

Explain shifts in equilibrium by applying Le Châtelier's principle.

Observe and **explain** the common-ion effect.

MATERIALS

- 24-well microplate
- CH_3COOH, 0.025 M
- CH_3COONa
- $CuSO_4$, 0.1 M
- $FeCl_3$, 0.025 M
- $Fe(NO_3)_3$
- glass stirring rod
- gloves
- HCl, 1.0 M
- KCl
- K_2HPO_4
- KSCN, 0.025 M
- lab apron
- methyl red indicator solution
- NH_3, 1.0 M
- NH_4SCN
- safety goggles
- test tube, 13 mm × 100 mm
- thin-stemmed pipets (8)

Always wear safety goggles, gloves, and a lab apron to protect your eyes and clothing. If you get a chemical in your eyes, immediately flush the chemical out at the eyewash station while calling to your teacher. Know the location of the emergency lab shower and eyewash station and the procedures for using them.

Equilibrium *continued*

Do not touch any chemicals. If you get a chemical on your skin or clothing, wash the chemical off at the sink while calling to your teacher. Make sure you carefully read the labels and follow the precautions on all containers of chemicals that you use. If there are no precautions stated on the label, ask your teacher what precautions to follow. Do not taste any chemicals or items used in the laboratory. Never return leftover chemicals to their original containers; take only small amounts to avoid wasting supplies.

Call your teacher in the event of a spill. Spills should be cleaned up promptly, according to your teacher's directions.

Acids and bases are corrosive. If an acid or base spills onto your skin or clothing, wash the area immediately with running water. Call your teacher in the event of an acid spill. Acid or base spills should be cleaned up promptly.

Procedure

1. Put on safety goggles, gloves, and a lab apron.

2. Put the 24-well plate on a sheet of white paper so the color changes will be more easily seen.

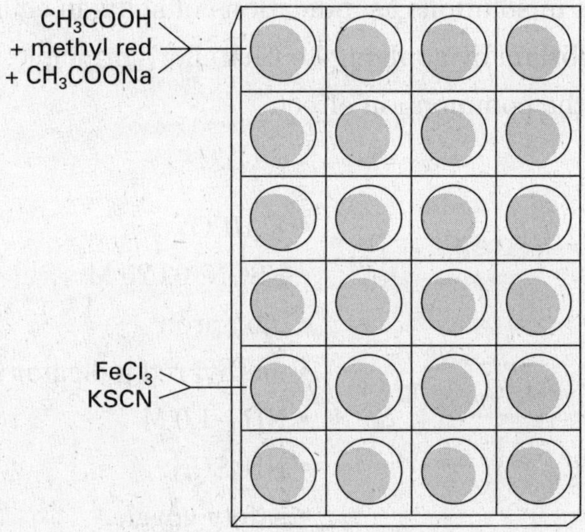

CH$_3$COOH
+ methyl red
+ CH$_3$COONa

FeCl$_3$
KSCN

Figure 1

PART 1–COMMON-ION EFFECT

3. Put 10 drops of 0.025 M acetic acid into a well in the top row of the plate. Add one drop of methyl red indicator solution. Mix the solution.

Observations: ____

4. Add one small crystal of CH$_3$COONa and mix to dissolve.

Observations: ____

Equilibrium *continued*

PART 2—LE CHÂTELIER'S PRINCIPLE

5. Put 3.0 mL of 0.1 M $CuSO_4$ in a test tube. The ion in the solution is $Cu(H_2O)_4^{2+}$.

Observations: ____

6. Add drops of 1.0 M NH_3 until the color changes and intensifies. The ion in the solution is $Cu(NH_3)_6^{2+}$.

Observations: ____

7. Add drops of 1.0 M HCl until the color changes again.

Observations: ____

____

8. Explain the solution's appearance after each addition in terms of Le Châtelier's principle.

____

____

____

____

PART 3—COMPLEX-ION EQUILIBRIUM

9. Mix 5 drops each of 0.025 M $FeCl_3$ and 0.025 M KSCN in a single well in the fifth row of the microplate. Mix thoroughly.

Observations: ____

10. Put 1 drop of the solution made in **step 9** into each of the four wells of the sixth row. Dilute each with 20 drops of distilled water. Add the following to the wells. Well *1*: a crystal of $Fe(NO_3)_3$; Well *2*: a crystal of NH_4SCN; Well *3*: a crystal of KCl; Well *4*: a crystal of K_2HPO_4.

Observations:

____

____

____

____

DISPOSAL

Clean all apparatus and your lab station. Return equipment to its proper place. Dispose of chemicals and solutions in the containers designated by your teacher. Do not pour any chemicals down the drain or in the trash unless your teacher directs you to do so. Wash your hands thoroughly before you leave the lab and after all work is finished.

Analysis

1. **Relating Ideas** Methyl red indicator has a pH range of 4.2 (red-violet color) to 6.2 (yellow color). Explain the color change that occurred in Part 1 of the procedure, when solid CH_3COONa was added to the acetic acid solution.

____

____

____

2. **Analyzing Information**

   a. What color is $Cu(H_2O)_4^{2+}$? _______________________________________

   b. What color is $Cu(NH_3)_6^{2+}$? _______________________________________

3. **Analyzing Information**

   a. When ammonia was added to the light-blue copper complex in **step 6,** what ion was formed?

   b. Write the balanced equation for this reaction.

   c. Use Le Châtelier's principle to explain why the addition of HCl caused the deep-blue solution to change to light blue.

____

____

____

____

4. **Relating Ideas** The Fe^{3+} ion and the SCN^- ion form the complex $FeSCN^{2+}$ ion, which is a deep-red color. Write the balanced equation for the formation of the $FeSCN^{2+}$ ion.

____

____

Equilibrium *continued*

5. Inferring Conclusions Indicate whether the product increased, decreased, or remained unchanged when the equilibrium conditions were changed in the following ways:

a. Fe^{3+} ions were added (well *1*).

b. SCN^- ions were added (well *2*).

c. KCl was added (well *3*).

____

____

____

Conclusions

1. Inferring Conclusions Some sunglasses darken when exposed to bright sunlight and become more transparent when they encounter shade. These sunglasses are made from a glass that contains small crystals of silver chloride. When photons of ultraviolet light hit these transparent crystals, the silver chloride changes into dark silver and chlorine atoms.

$$AgCl \xrightleftharpoons{\text{light}} Ag^0 + Cl^0$$

In the shade, the equilibrium reverses, producing transparent silver chloride crystals again. This equilibrium, however, does not occur in solutions. When a freshly prepared precipitate of silver chloride is exposed to sunlight, the surface turns black. Although this is the same reaction as in the photosensitive sunglasses, this time it is not reversible. Explain why. (Hint: How are the environments of the products different in the two situations?)

____

____

____

2. Analyzing Conclusions Many changes took place in the three parts of this experiment. Write a single statement to summarize all of these changes.

____

____

____

(Microscale)

Oxidation-Reduction Reactions

A substance that loses electrons during a chemical reaction is said to be oxidized. A substance that gains electrons is said to be reduced. If one reactant gains electrons, another must lose an equal number. Thus, oxidation and reduction must occur simultaneously and to a comparable degree.

The stronger the tendency of a species to take electrons, the greater is its strength as an oxidizing agent and the more easily it is reduced. The stronger the tendency of a species to give up electrons, the greater is its strength as a reducing agent and the more readily it is oxidized. The silver ion, Ag^+, has a strong tendency to acquire an electron to form the silver atom, Ag. Thus, the Ag^+ ion is a strong oxidizing agent. In this experiment, you will determine the relative strengths of some metals as reducing agents and the relative strengths of their ions as oxidizing agents.

OBJECTIVES

• **Observe** oxidation-reduction reactions between metals.

• **Describe** typical oxidation-reduction reactions.

• **Determine** relative strengths of some oxidizing and reducing agents.

MATERIALS

- 24-well microplate
- thin-stemmed pipets, 6
- 0.1 M $Cu(NO_3)_2$
- 0.1 M $FeCl_3$
- 0.1 M $KMnO_4$
- 0.1 M $Mg(NO_3)_2$
- 0.1 M $Zn(NO_3)_2$

- 1.0 M H_2SO_4
- 0.5 cm copper wire, 3 pieces
- iron(II) sulfate
- 5 cm magnesium ribbon
- 0.5 cm zinc strip, 3 pieces
- tin(II) chloride

Always wear safety goggles and a lab apron to protect your eyes and clothing. If you get a chemical in your eyes, immediately flush the chemical out at the eyewash station while calling to your teacher. Know the locations of the emergency lab shower and eyewash and the procedures for using them.

Do not touch any chemicals. If you get a chemical on your skin or clothing, wash the chemical off at the sink while calling to your teacher. Make sure you carefully read the labels and follow the precautions on all containers of chemicals that you use. If there are no precautions stated on the label, ask your teacher what precautions you should follow. Do not taste any chemicals or items used in the laboratory. Never return leftovers to their original containers; take only small amounts to avoid wasting supplies.

Oxidation-Reduction Reactions *continued*

 **Call your teacher in the event of a spill.** Spills should be cleaned up promptly, according to your teacher's directions.

PROCEDURE

1. Place five drops of copper(II) nitrate solution, $Cu(NO_3)_2$, in each of the four wells in the top row of a microplate, as shown in **Figure A.** Place five drops of zinc nitrate, $Zn(NO_3)_2$, in each of the four wells in the next row. Place five drops of magnesium nitrate, $Mg(NO_3)_2$, in each of the four wells in the third row. Record the original color of each solution in **Data Table 1.**

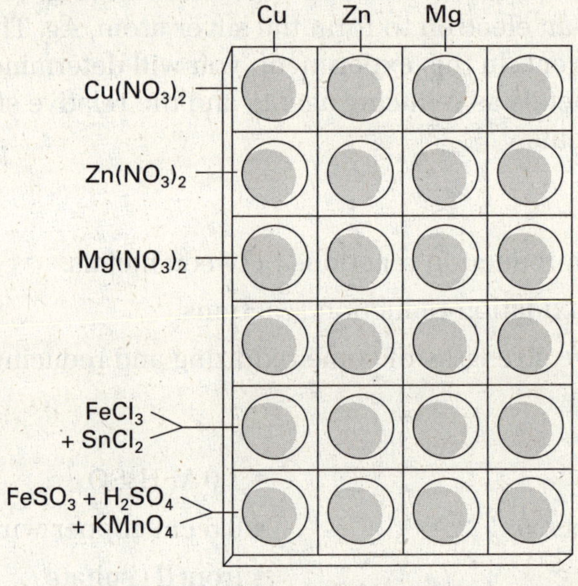

Figure A

2. Add a small piece of copper wire to each of the three solutions in the first column of the microplate. Add a small piece of zinc to the three solutions in the second column of the plate. Add a small piece of magnesium to the three solutions in the third column. The fourth column will be used for comparison.

3. Put the microplate on a white sheet of paper and observe for 10–15 min. Record the results in **Data Table 1.** Use *NR* for no reaction.

4. Put five drops of 0.1 M iron(III) chloride, $FeCl_3$, in one of the empty wells. Add one small crystal of tin(II) chloride, $SnCl_2$, to this solution. Record the results in **Data Table 2.**

5. Add a small crystal of iron(II) sulfate, $FeSO_3$, to another of the empty wells. Add 10 drops of water and five drops of 1.0 M sulfuric acid, H_2SO_4. Add 0.1 M potassium permanganate, $KMnO_4$, dropwise, mixing after each addition. Continue adding $KMnO_4$ until two color changes have occurred. Record the results in **Data Table 2.**

Oxidation-Reduction Reactions *continued*

DISPOSAL

6. Clean all apparatus and your lab station. Return equipment to its proper place. Dispose of chemicals and solutions in the containers designated by your teacher. Do not pour any chemicals down the drain or in the trash unless your teacher directs you to do so. Wash your hands thoroughly before you leave the lab and after all work is finished.

DATA TABLE 1

	Alone	Cu	Zn	Mg
$Cu(NO_3)_2$				
$Zn(NO_3)_2$				
$Mg(NO_3)_2$				

DATA TABLE 2

	Color of Iron Compound	Color of Solution
$FeCl_3 + SnCl_2$		
$FeSO_3 + H_2SO_4 + KMnO_4$		

Analysis

Refer to **Data Table 1** to answer questions 1–6.

1. Organizing Data Which metal was oxidized by two other ions?

____

2. Organizing Data Which metal was oxidized by only one other ion?

____

3. Organizing Data Which metal was not oxidized by any of the ions?

____

4. Analyzing Results Arrange the three metals in order of their relative strengths as reducing agents, placing the strongest first.

____

Oxidation-Reduction Reactions *continued*

5. Analyzing Results Arrange the three metallic ions in order of their relative strengths as oxidizing agents, placing the strongest first. Write the reduction half-reaction for each ion.

____

____

____

6. Inferring Results Copper is oxidized in the presence of silver ions. The net ionic reaction is the following:

$$Cu + 2Ag^+ \rightarrow Cu^{2+} + 2Ag$$

Write net ionic equations for the following: (Hint: Use your answers to Questions **4** and **5** to determine which metal is oxidized and which ion is reduced.)

a. the reaction of copper and zinc

____

b. the reaction of zinc and magnesium

____

c. the reaction of copper and magnesium.

____

7. Analyzing Results In step **4**, the Fe^{3+} ion was reduced to the Fe^{2+} ion.

a. What was the reducing agent?

____

b. What change did the Sn^{2+} ion undergo?

____

c. Write the net ionic equation for the overall reaction:

$$2Fe^{3+} + 6Cl^- + Sn^{2+} + 2Cl^- \rightarrow 2Fe^{2+} + 4Cl^- + Sn^{4+} + 4Cl^-$$

____

8. Analyzing Results The permanganate ion, MnO_4^-, which is purple in color, is a strong oxidizing agent. The manganese(II) ion, Mn^{2+}, is practically colorless. What occurred during the addition of potassium permanganate to the Fe^{2+} ions in step **5**?

____

____

____

Oxidation-Reduction Reactions *continued*

Conclusions

1. Predicting Outcomes What would happen to the metals if iron nails were used to secure sheets of copper to a roof?

____

____